AF595105

Biocatalytic Synthesis of Bioactive Compounds

Biocatalytic Synthesis of Bioactive Compounds

Editor

Josefina Aleu

MDPI • Basel • Beijing • Wuhan • Barcelona • Belgrade • Manchester • Tokyo • Cluj • Tianjin

Editor
Josefina Aleu
University of Cádiz
Spain

Editorial Office
MDPI
St. Alban-Anlage 66
4052 Basel, Switzerland

This is a reprint of articles from the Special Issue published online in the open access journal *Molecules* (ISSN 1420-3049) (available at: https://www.mdpi.com/journal/molecules/special_issues/Biocatalytic_Synthesis).

For citation purposes, cite each article independently as indicated on the article page online and as indicated below:

LastName, A.A.; LastName, B.B.; LastName, C.C. Article Title. *Journal Name* **Year**, *Article Number*, Page Range.

ISBN 978-3-03943-571-5 (Hbk)
ISBN 978-3-03943-572-2 (PDF)

Contents

About the Editor

Josefina Aleu studied Chemistry at the University of Cádiz, where she graduated in 1991. After working with Prof. J. R. Hanson at Sussex University for several months in 1995, she then completed her Ph.D. degree in 1996 under the supervision of Prof. Collado on the design of selective fungicides. She then worked for 15 months as a postdoctoral fellow with Prof. C. Fuganti at Politecnico di Milano (Italy) on the biocatalytic synthesis of flavors. She is currently Senior Lecturer of Organic Chemistry at the University of Cádiz, and her current research interests are directed towards the biocatalytic synthesis of bioactive molecules, design of selective fungicides, and studies of the metabolism of marine microorganisms.

Preface to "Biocatalytic Synthesis of Bioactive Compounds"

Biocatalysis, the application of enzymes as catalysts for chemical synthesis, has become an increasingly valuable tool for the synthetic chemist. Enzymatic transformations carried out by partially purified enzymes or whole-cell catalysts are used for the production of a wide variety of compounds ranging from bulk to fine chemicals. In the field of fine chemicals, the main application clearly lies in the exploitation of the outstanding properties of these biocatalysts with respect to chemoselectivity, regioselectivity, and especially stereoselectivity in the production of enantiomerically pure compounds. Until recently, chemical development companies regarded biocatalysis as a method to be attempted only when all other chemical options failed. Currently, there are clear signs that this view is changing radically, with many of the new process developments revealing the benefits to be gained by using biocatalysis on a commercial scale. Biocatalytic processes have a number of advantages over the corresponding chemical methods. The conditions for such processes are mild and, in the majority of cases, do not require the protection of other functional groups. Moreover, eliminating the need for several chemical synthetic steps has had a dramatic impact on the overall economics. In many cases, biological methods are also enantiospecific, allowing for the production of chiral products from racemic mixtures. Furthermore, the features governing their regiospecificity differ from those controlling chemical specificity and, indeed, it is possible to obtain biotransformations at centers that are chemically unreactive. Economically, biocatalytic processes are often cheaper and more direct than their chemical counterparts, and the conversions normally proceed under conditions that are regarded as ecologically acceptable.

The aim of this Special Issue is to present studies focused on biocatalysis as applied to the organic synthesis of bioactive compounds and their precursors. The obtained contributions deal with biotransformations, including the stereoselective synthesis of bioactive chemical compounds, active pharmaceutical ingredients, and natural products. The production of interesting bioactive compounds is covered in the eight articles comprising seven research articles and a review.

Two of these papers deal with fungal metabolism and fungi-mediated biotransformations. The first paper deals with improving β-pinene bioconversion by mutagenesis and adaptation of the fungus *Chrysosporium pannorum* A-1. A second paper describes a study on the biotransformation of bioactive coumarins by filamentous fungi.

On the other hand, the conversion of oleic acid into azelaic and pelargonic acid by a chemoenzymatic route represents a sustainable alternative to ozonolysis, currently employed at the industrial scale to perform the reaction. In addition, the findings on a gram-scale limonene production process using engineered Escherichia coli provide a basis for the development of an economic and industrially relevant bioprocess.

Alkene cleavage is a possibility to generate aldehydes with olfactory properties for the fragrance and flavor industry. A dye-decolorizing peroxidase (DyP) of the basidiomycete *Pleurotus sapidus* is the first described DyP with alkene cleavage activity towards aryl alkenes and showed potential as biocatalyst for flavor production.

Isoflavones in soybeans are well-known phytoestrogens. Soy isoflavones present in conjugated forms are converted to aglycone forms during processing and storage. Isoflavone aglycones (IFAs) of soybeans in human diets have poor solubility in water, resulting in low bioavailability and bioactivity. Enzyme-mediated glycosylation is an efficient and environmentally friendly way to modify the physicochemical properties of soy IFAs. Thus, Jung et al. studied the enrichment of polyglucosylated isoflavones from soybean isoflavone aglycones using optimized amylosucrase transglycosylation. In the same line, Chang et al. studied the potential industrial production of a well-soluble, alkaline-stable, and anti-inflammatory isoflavone glucoside from 8-hydroxydaidzein glucosylated by recombinant amylosucrase of *Deinococcus geothermalis*.

Finally, the review discusses the use of halogenating enzymes in fine chemical syntheses, detailing the many steps that are still needed before halogenating enzymes are considered reliable, flexible, and sustainable catalysts for halogenation.

Overall, these eight contributions provide the reader with relevant fresh insights into the use of enzymes and whole cells as biocatalysts.

Josefina Aleu

Editor

Article

Potential Industrial Production of a Well-Soluble, Alkaline-Stable, and Anti-Inflammatory Isoflavone Glucoside from 8-Hydroxydaidzein Glucosylated by Recombinant Amylosucrase of *Deinococcus geothermalis*

Te-Sheng Chang [1,†,*], Tzi-Yuan Wang [2,†], Szu-Yi Yang [1], Yu-Han Kao [1], Jiumn-Yih Wu [3,*] and Chien-Min Chiang [4,*]

1 Department of Biological Sciences and Technology, National University of Tainan, Tainan 70005, Taiwan; szuyi08231995@gmail.com (S.-Y.Y.); aa0920281529@gmail.com (Y.-H.K.)
2 Biodiversity Research Center, Academia Sinica, Taipei 115, Taiwan; tziyuan@gmail.com
3 Department of Food Science, National Quemoy University, Kinmen County 892, Taiwan
4 Department of Biotechnology, Chia Nan University of Pharmacy and Science, No. 60, Sec. 1, Erh-Jen Rd., Jen-Te District, Tainan 71710, Taiwan
* Correspondence: mozyme2001@gmail.com (T.-S.C.); wujy@nqu.edu.tw (J.-Y.W.); cmchiang@mail.cnu.edu.tw (C.-M.C.); Tel.: +886-6-2606283 (T.-S.C.); +886-82-313310 (J.-Y.W.); +886-6-2664911 (ext. 2542) (C.-M.C.); Fax: +886-6-2606153 (T.-S.C.); +886-82-313797 (J.-Y.W.); +886-6-2662135 (C.-M.C.)
† These authors contributed equally to this work.

Academic Editor: Josefina Aleu
Received: 30 May 2019; Accepted: 14 June 2019; Published: 15 June 2019

Abstract: 8-Hydroxydaidzein (8-OHDe), an *ortho*-hydroxylation derivative of soy isoflavone daidzein isolated from some fermented soybean foods, has been demonstrated to possess potent anti-inflammatory activity. However, the isoflavone aglycone is poorly soluble and unstable in alkaline solutions. To improve the aqueous solubility and stability of the functional isoflavone, 8-OHDe was glucosylated with recombinant amylosucrase of *Deinococcus geothermalis* (DgAS) with industrial sucrose, instead of expensive uridine diphosphate-glucose (UDP-glucose). One major product was produced from the biotransformation, and identified as 8-OHDe-7-α-glucoside, based on mass and nuclear magnetic resonance spectral analyses. The aqueous solubility and stability of the isoflavone glucoside were determined, and the results showed that the isoflavone glucoside was almost 4-fold more soluble and more than six-fold higher alkaline-stable than 8-OHDe. In addition, the anti-inflammatory activity of 8-OHDe-7-α-glucoside was also determined by the inhibition of lipopolysaccharide-induced nitric oxide production in RAW 264.7 cells. The results showed that 8-OHDe-7-α-glucoside exhibited significant and dose-dependent inhibition on the production of nitric oxide, with an IC_{50} value of 173.2 μM, which remained 20% of the anti-inflammatory activity of 8-OHDe. In conclusion, the well-soluble and alkaline-stable 8-OHDe-7-α-glucoside produced by recombinant DgAS with a cheap substrate, sucrose, as a sugar donor retains moderate anti-inflammatory activity, and could be used in industrial applications in the future.

Keywords: 8-hydroxydaidzein; stable; soluble; anti-inflammation; amylosucrase; *Deinococcus geothermalis*

---

1. Introduction

Daidzin (D) and genistin (G) are major components of soy isoflavone in soybeans, which are the glucoside derivatives of soy isoflavone aglycone daidzein (De) and genistein (Ge), respectively. De and

Ge aglycones have been studied widely and demonstrated multiple bioactivities [1]. In addition to De and Ge, some soy isoflavone derivatives have been isolated from fermented soybean foods, and shown to possess multiple bioactivities [2]. 8-Hydroxydaidzein (8-OHDe), an *ortho*-hydroxylation derivative of De, was first isolated from the fermentation broth of *Streptomyces* sp. In the presence of soybeans [3], and then from many fermented soybean foods [4–8]. In recent decades, 8-OHDe has been evaluated to possess many bioactivities, such as suppression of multi-drug resistance in Caco-2 colon adenocarcinoma cells [9], irreversible inhibition of tyrosinase [10,11], anti-melanogenesis [12,13], inhibition of aldose reductase [14], and anti-inflammation [15,16].

Recently, Seo et al. and Wu et al. developed mass production processes for 8-OHDe from biotransformation of De by *Aspergillus oryzae* [17,18]. The availability of a large quantity of 8-OHDe provides more opportunities for the application of 8-OHDe in the industry. However, although 8-OHDe has many bioactivities, and can be obtained on a large scale, the isoflavone has drawbacks of low solubility and high instability in alkaline solutions [19,20]. These drawbacks limit the applications of 8-OHDe, unless one can improve the half-life of isoflavone with higher solubility and stability.

Biotransformation of natural products by microorganisms and/or enzymes provides a route to improve the properties of the original compounds [21,22]. Among different kinds of flavonoid biotransformation, glycosylation of flavonoids usually holds great promise to increase the solubility of the original compounds. For example, the aqueous solubility of soy isoflavones is improved about 30-fold through glycosylation [23]. Likewise, the glycosyl-biotransformation of 8-OHDe might improve its aqueous solubility and stability. In nature, glycosylation of flavonoids is usually catalyzed with glycosyltransferases (GTs), which use activated uridine diphosphate-glucose (UDP-glucose) as a sugar donor, and transfer the sugar to a flavonoid acceptor [24]. A previous study used the recombinant BsGT110 from *Bacillus subtilis* to catalyze glucosylation of 8-OHDe [20]. The results showed that the aqueous solubility and stability of the isoflavone glucosides (8-OHDe-7-*O*-β-glucoside and 8-OHDe-8-*O*-β-glucoside) were greatly improved. In addition, such biotransformation was not easily scaled up to the industrial level because of the expensive substrate, UDP-glucose.

However, some glycoside-hydroxylases (GHs) also catalyze glucosylation of flavonoids [25,26]. For example, the amylosucrase, GH of the GH13 family, is able to catalyze glucosylation of flavonoids with cheaper sucrose as a sugar donor, which is one-millionth the cost of UDP-glucose [26]. An amylosucrase from *Deinococcus geothermalis* (DgAS) is one of the promising bio-catalysts in glucosylation of phenolic molecules, because of its thermally-stable and higher activity than other amylosucrases [27,28]. In the present study, the DgAS enzyme was produced in recombinant *Escherichia coli,* and the purified DgAS was detected to catalyze the glucosylation of 8-OHDe. The biotransformed glucosidic product was then purified with chromatography, identified with spectrometric methods. The aqueous solubility, stability, and anti-inflammatory assay of the produced isoflavone glucoside were determined.

2. Results

2.1. Production of DgAS Protein in Recombinant Escherichia coli

The *DgAS* gene was amplified from genomic DNA of *Deinococcus geothermalis*, and subcloned into an expression plasmid *pETDuet-1* fused with six histidine residues in the N-terminal. The constructed *pETDuet-DgAS* plasmid (Figure 1a) was overexpressed in *E. coli*, and induced with 0.2 mM of isopropyl β-D-1-thiogalactopyranoside (IPTG). The soluble DgAS proteins were then successfully purified with Ni^{2+} chelate affinity chromatography, as shown in sodium dodecyl sulfate polyacrylamide gel electrophoresis (SDS-PAGE) (Figure 1b).

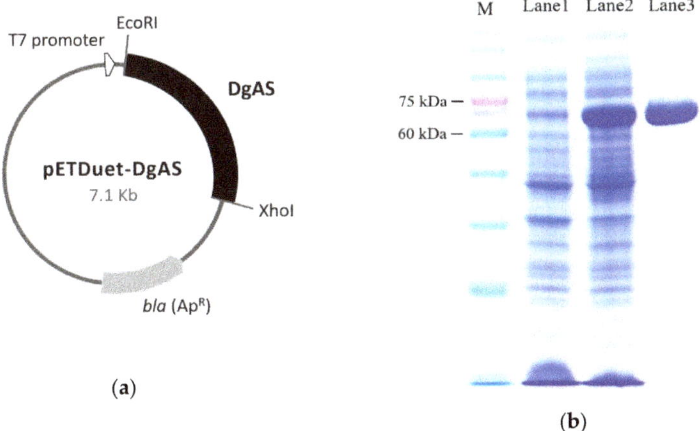

Figure 1. Production of the DgAS from *Deinococcus geothermalis* in *E. coli*. (**a**) Diagram of the constructed plasmid; (**b**) SDS-PAGE of the produced DgAS in recombinant *E. coli*. M: protein marker; lane 1: total protein without IPTG-induction; lane 2: total protein with IPTG-induction for 20 h; lane 3: purified DgAS.

2.2. Biotransformation of 8-OHDe by Recombinant DgAS Protein

The purified DgAS were used to catalyze the biotransformation of 8-OHDe. The reaction contained 0.1 mg/mL of 8-OHDe, 12.5 μg/mL of DgAS, and 20 mM of sucrose as the sugar donor. The reaction mixtures at 0 min (the dashed line) and after 30-min incubation were analyzed with ultra-performance liquid chromatography (UPLC). One major product with a retention time (RT) of 3.1 min was observed (Figure 2).

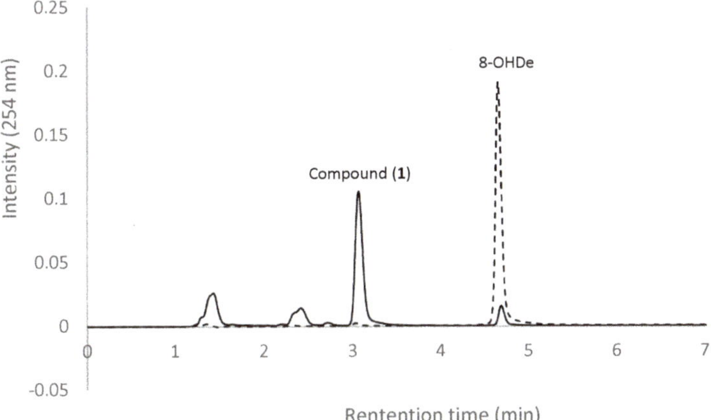

Figure 2. Biotransformation of 8-OHDe with DgAS. The reaction mixtures at 0 min (dashed line) and 30 min (solid line) were analyzed with UPLC. The UPLC operation conditions are described in Materials and Methods.

To optimize compound (**1**) production for further analysis, a standard mixture with 1 mg/mL of 8-OHDe and 125 μg/mL of DgAS was carried out at different temperatures, pH levels, and serial concentrations of sucrose for 30-min incubation. After incubation, the production conversion of the major product was determined with UPLC (Figure 3). The results revealed the optimal condition for the production of compound (**1**) from 8-OHDe by the recombinant DgAS is pH 7, 40 °C, and 300 mM of sucrose.

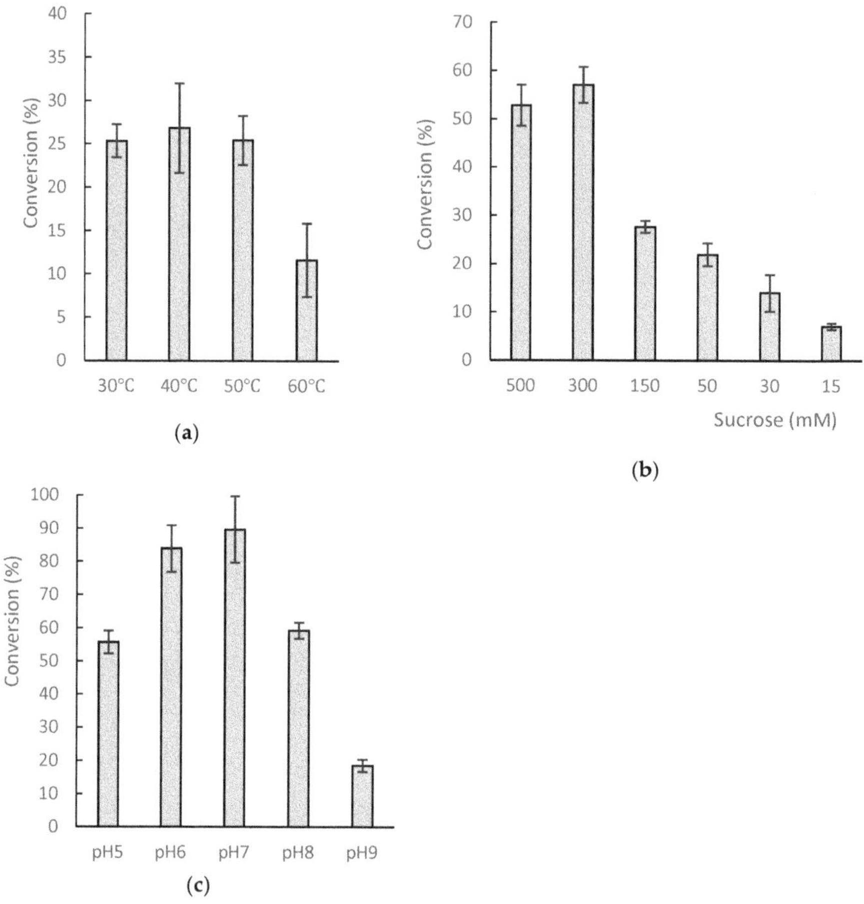

Figure 3. Optimal condition for the production of compound (**1**) from 8-OHDe by DgAS. (**a**) 125 μg of the purified DgAS and 1 mg/mL of 8-OHDe were incubated at different temperatures, (**b**) serial concentrations of sucrose, and (**c**) pH levels for 30 min. After incubation, the reaction was analyzed with UPLC. The conversion was calculated by dividing the amount of the produced compound (**1**) in each reaction by the theoretical production value (1.6 mg) for 100% conversion. The mean (n = 3) is shown, and the standard deviations are represented by error bars.

2.3. Identification of the Major Product

To resolve the chemical structure of the product, the biotransformation was scaled up to 100 mL, with 1 mg/mL of 8-OHDe, 125 μg/mL of DgAS, and 300 mM of sucrose at pH 7 and 40 °C for 30-min incubation. About 90 mg of the product in the 100-mL reaction was purified with preparative high-performance liquid chromatography (HPLC). Based on the value of the optimal conversion (89.3%) (Figure 3), the maximum production yield of compound (**1**) from 100 mg of 8-OHDe is 142.3 mg (160 mg × 0.89); thus, the purification recovery yield is 62.5% (90/142). The chemical structure of the purified compound (**1**) was resolved with mass and nuclear magnetic resonance (NMR) spectral analysis. The mass analysis of the compound showed an $[M + H]^+$ ion peak at *m/z*: 433.18 in the electrospray ionization mass (ESI-MS) spectrum corresponding to the molecular formula $C_{21}H_{20}O_{10}$. Then ^1H and

^{13}C nmR spectra, including distortionless enhancement by polarization transfer (DEPT), heteronuclear single quantum coherence (HSQC), heteronuclear multiple bond connectivity (HMBC), correlation spectroscopy (COSY), and nuclear Overhauser effect spectroscopy (NOESY) spectra, were obtained, and the ^{1}H and ^{13}C nmR signal assignments were conducted accordingly (shown in Figures S1–S7). In addition to the signals of 8-OHDe, seven proton signals (from 3.22 to 5.48 ppm) and six carbon signals (from 60–100 ppm) indicated a glucose moiety. The J coupling constant (3.6 Hz) of the anomeric proton (5.48 ppm) in the ^{1}H nmR spectra of compound (1) indicated the α-configuration for the glucopyranosyl moiety. The cross peak of H-1″ with C-7 (5.48/148.8 ppm) in the HMBC spectrum, as well as the cross peak of H-1″ with H-6 (5.48/7.38 ppm) in the NOESY spectrum, demonstrated the structure of compound (**1**) was 8-OHDe-7-*O*-α-glucoside. The key HMBC and NOESY correlations of compound (**1**) are shown in Figure S8, and the spectroscopic data is listed in Table S1. The downfield shift of the ^{1}H signal H-1″, the anomeric proton, of compound (**1**) compared to that of 8-OHDe-7-*O*-β-glucoside [20], indicated their different microenvironments. Figure 4 illustrates the biotransformation process of 8-OHDe by DgAS.

HO OH O O OH
DgAS
Sucrose
HO OH HO OH O O OH O O OH
8-OHDe
8-OHDe 7-O-α-glucoside

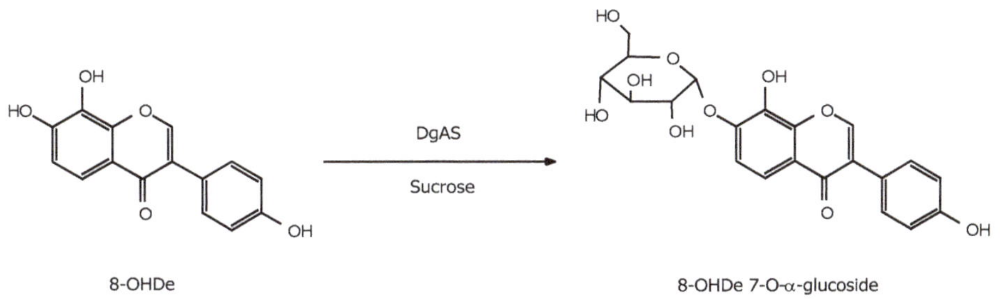

Figure 4. Biotransformation process of 8-OHDe by DgAS.

2.4. Solubility, Stability, and Anti-Inflammatory Activity of 8-OHDe 7-α-glucoside.

The aqueous solubility of 8-OHDe-7-*O*-α-glucoside was examined (Table 1). The results revealed that the maximum aqueous solubility of the 8-OHDe-7-*O*-α-glucoside is 3.92-fold, higher than that of 8-OHDe.

Table 1. Aqueous solubility of 8-OHDe and 8-OHDe-7-*O*-α-glucoside.

Compound	Aqueous-Solubility (mg/L)	Fold
8-OHDe	47.3	1
8-OHDe-7-*O*-α-glucoside	185.4	3.92

In addition, 8-OHDe is very unstable in the alkaline condition [19]. This property shortens the storage time of 8-OHDe in cosmetic or pharmaceutical products, and limits applications of 8-OHDe. Thus, the stability of the 8-OHDe-7-*O*-α-glucoside and 8-OHDe was compared (Figure 5). The half-time of 8-OHDe was 15.8 h, and only 6.8% of 8-OHDe remained in 50 mM of Tris buffer (pH 8.0) after 96-h incubation at 20 °C. However, 94.6% of 8-OHDe-7-*O*-α-glucoside still remained after 96-h incubation at the same condition. The half-time of 8-OHDe-7-*O*-α-glucoside was much longer than 96 h. Thus, 8-OHDe-7-*O*-α-glucoside is more than six-fold more stable than 8-OHDe in an alkaline solution.

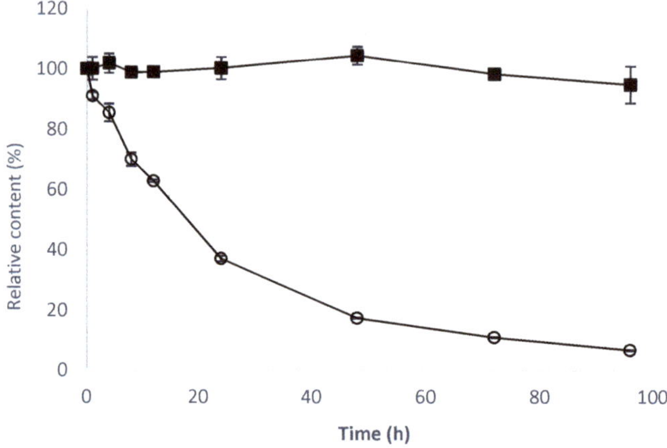

Figure 5. Alkaline stability of 8-OHDe (open circle) and 8-OHDe-7-*O*-α-glucoside (closed square). A total of 1 mg/mL of the tested compound was dissolved in 50 mM of Tris buffer at pH 8.0, and stored at 20 °C for 96 h. During the storage time, samples were taken out for UPLC at the determined interval times. The mean (n = 3) is shown, and the standard deviations are represented by error bars.

Since the 8-OHDe-7-*O*-α-glucoside possesses much higher aqueous solubility and alkaline stability than those of 8-OHDe, and 8-OHDe was recently demonstrated with potent anti-inflammatory activity [15,16], the anti-inflammatory activity of 8-OHDe-7-*O*-α-glucoside and 8-OHDe was determined by the inhibition ability on lipopolysaccharide (LPS)-induced nitric oxide (NO) production in murine macrophage RAW264.7 cells. Macrophages are involved in chronic inflammation, and once macrophages are elicited, an inflammatory mediator, such as NO, is produced. Thus, the NO level in the culture supernatant could be measured as an index of inflammatory mediators. The results of the anti-inflammatory assays indicated 8-OHDe-7-α-glucoside exhibited statistically significant and dose-dependent inhibitory activity with an IC_{50} value of 173.2 ± 12.9 μM (Figure 6a), while 8-OHDe showed potent anti-inflammatory activity with an IC_{50} value of 34.5 ± 5.3 μM (Figure 6c). In addition, the results of cell survival assay indicated that 100 ng/mL of LPS treatment would not induce statistically significant cell death as the false-positive signal and reduction of NO by the tested isoflavones was not due to the cytotoxicity of the isoflavones (Figure 6b,d).

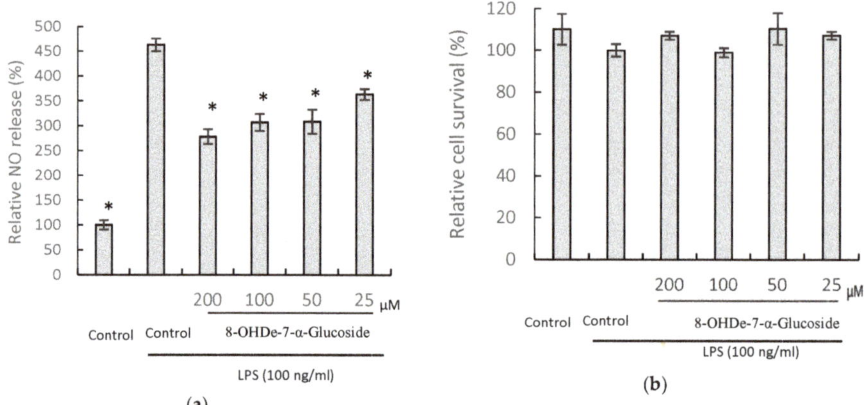

Figure 6. *Cont.*

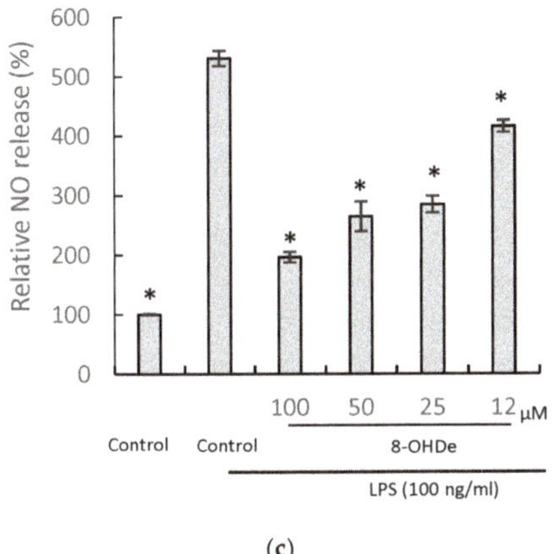

(c)

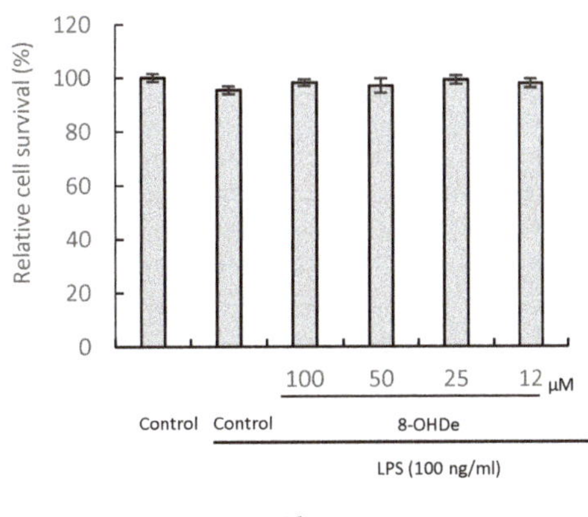

(d)

Figure 6. Effects of 8-OHDe-7-α-glucoside (**a**,**b**) and 8-OHDe (**c**,**d**) on the inhibition of LPS-induced NO production (**a**,**c**) and cell survival (**b**,**d**) in murine macrophage RAW264.7 cells. Cells were incubated with the indicated concentrations of isoflavone for 1 h before treatment with LPS (100 ng/mL) for 24 h. The amounts of NO were determined using the Griess reagent in the culture medium. Cell viability was determined with MTT assay. Each value indicates the mean ± standard deviation (SD), and is representative of the results obtained from four independent experiments. * ($p < 0.001$) is statistically significantly different from the value for the cells treated with LPS treatment alone.

3. Discussion

Most GTs require additional expensive sugar donors, such as UDP-glucose, to generate glucose conjugates. In contrast, amylosucrase is particularly useful for glucosylation of flavonoids, due to its ability to use sucrose, an inexpensive and abundant renewable substrate, as a sugar donor. The studied amylosucrase (DgAS) was first identified by Stephane et al. [27], who expressed DgAS with N-terminal fusion of glutathione-S-transferase (GST) in *E. coli*, and found that the GST-DgAS was an inactive protein insoluble in the inclusion bodies. The N-terminal GST needed to be removed for the functional glucosylation activity. In contrast, Lee et al. recently produced recombinant DgAS in *E. coli* with fusion of His-tag in its C-terminal, and successfully purified the fusion DgAS as an active form with a simple Ni^{2+} affinity chromatography method [29]. The authors used the DgAS to catalyze glucosylation of hydroquinone forming α-arbutin, and their results showed an optimal reaction condition at pH 7, 40 °C, and 300 mM of sucrose. In the present study, we expressed DgAS with N-terminal His-tag fusion in *E. coli*, and successfully purified the fusion DgAS as an active form with a similar method (Figures 1 and 2). After induction, 12.5 mg of well-soluble DgAS could be purified from 150 mL of the cell cultivation. Moreover, the optimal condition of major compound production by the DgAS enzyme (Figure 3) was consistent with that of Lee et al. [29], although we used 8-OHDe as the sugar acceptor. Thus, it seems that either N-terminal (the present study) or C-terminal [29] His-tag fusion is a more suitable strategy than GST fusion for DgAS production in *E. coli*.

Based on the naturally catalytic property of DgAS, which syntheses glucan polymer (amylose) from hydrolysis of the sucrose substrate, DgAS prefers catalyzing glucosylation sites at the hydroxyl group of an existing sugar moiety as the sugar acceptor [27,28]. Thus, it is easy to predict the catalytic products from the flavonoid glycosides substrate by DgAS, which would add a glucose moiety from sucrose to the hydroxyl group of sugar moiety in the flavonoid glycoside [30,31]. However, it is difficult to predict the *O*-glucosylation product when a flavonoid aglycone contains multiple hydroxyl groups as the sugar acceptor, due to the lack of sugar moiety in the structure. Thus far, DgAS has been reported to catalyze *O*-glucosylation only toward three flavonoid aglycones, catechin (5,7,3′,4′-tetrahydroxyflava-3-ol) [32], baicalein (5,6,7-trihydroxyflavone) [33], and 8-OHDe (7,8,4′-trihydroxyisoflavone, the present study), to form catechin-3′-*O*-α-glucoside, baicalein-6-*O*-α-glucoside, and 8-OHDe-7-*O*-α-glucoside, respectively. The three flavonoid aglycones contain a hydroxyl group at C7; however, only 8-OHDe was glucosylated at the C7 site by DgAS (Figure 4). It seems that DgAS prefers the two sites C3′ and C6 rather than C7 on flavonoid aglycones for glucosylation. The detail molecular mechanism for the glucosylation sites toward flavonoid aglycones by DgAS must be systematically studied in the future.

Drugs with poor aqueous solubility exhibit dilution rate-limited absorption in the membrane of the gastrointestinal tract; therefore, enhancing the solubility of drugs that have poor water solubility is an important issue in pharmaceutical research [32]. For example, although 8-OHDe is highly valuable in pharmaceutical research [3–16], its applications have been restricted due to its poor water solubility and alkaline instability in aqueous solution. Furthermore, the stability of flavonoids at various pH levels is also important for absorption in the gut, because of the sharp increase in pH from the acidic stomach to the slightly alkaline intestine. Glucosylation of flavonoids could improve such limitations. The stability and solubility of the glycosylated product were found to be drastically increased when compared to their aglycones [20]. In the present study, glucosylation significantly extended the half-life of 8-OHDe (Figure 5), and increased 3.92-fold the aqueous solubility of 8-OHDe (Table 1). Therefore, the glucosylated product, 8-OHDe-7-*O*-α-glucoside, has high potential in the pharmaceutical industry.

However, glucosylated flavonoids sometimes lose the bioactivity of their flavonoid precursors, although glucosylation could improve aqueous solubility, and extend the half-life of flavonoids. Since 8-OHDe has been shown to exhibit high anti-inflammatory activity [15,16], we wanted to determine the effect of glucosylated 8-OHDe on anti-inflammatory activity. As expected, the results showed that 8-OHDe-7-*O*-α-glucoside maintains only 20.1% of the anti-inflammatory activity of 8-OHDe (Figure 6), which is consistent with Hamalainen et al.'s results [34]. Hamalainen et al. also found inhibition of NO production in RAW264.7 cells by soy isoflavone genistein (97% of inhibition of NO production at the 100-μM concentration) was about five-fold higher than that of genistein-7-β-glucoside, genistin (17% of inhibition of NO production at the 100-μM concentration). The results reveal that the glucosyl group of the isoflavone skeleton would induce a negative effect on the exhibition of the anti-inflammatory activity. The detailed structure-activity relationship needs to be studied in the future. In addition, the signaling pathways involving in the anti-inflammatory activity by 8-OHDe in macrophage cells was already determined by Wu et al. [15] and Kim et al. [16], who demonstrated that the signals in both nuclear factor κB (NF- κB) and activator protein 1 (AP1) signaling pathways were inhibited by 8-OHDe in the anti-inflammation. Therefore, it is worthy to know if 8-OHDe-7-α-glucoside has similar signaling mechanism of anti-inflammation in the future. Nevertheless, the trade-off improved solubility, and long half-life alkaline-stability could extend the potential application of 8-OHDe-7-*O*-α-glucoside in anti-inflammation.

4. Materials and Methods

4.1. *Microorganisms, Animal Cells, and Chemicals*

Deinococcus geothermalis DSM11300 (BCRC17378) and mouse macrophage cells RAW 264.7 (BCRC60001) were obtained from the Bioresources Collection and Research Center (BCRC, Food Industry Research and Development Institute, Hsinchu, Taiwan), and cultivated according to the BCRC protocol. *E. coli* BL21 (DE3) and the *pET-Duet-1* expression plasmid were obtained from the Novagen Company (Madison, WI, USA). Restriction enzymes and DNA modified enzymes were obtained from New England Biolabs (Ipswich, MA, USA). 8-OHDe was prepared according to Dr. Wu's method [18]. IPTG, 3-(4,5-dimethylthiazol-2-yl)-2,5-diphenyltetrazolium bromide (MTT), dimethyl sulfoxide (DMSO), LPS, and Greiss reagent were purchased from Sigma (St. Louis, MO, USA). The other reagents and solvents used are commercially available.

4.2. Preparation of Recombinant DgAS Enzyme

DgAS was amplified from the genome of *Deinococcus geothermalis* with polymerase chain reaction (PCR) with a primer set: forward 5′-cccgaattcgCTGAAAGACGTGCTCACTTCTGAAC-3′ and reverse 5′-aaactcgagTTATGCTGGAGCCTCCCCGGCGGTC-3′, which contained EcoRI and XhoI restriction sites, respectively. The amplified DNA fragment (1.95 kb) was digested with EcoRI and XhoI, and ligated into the corresponding sites on the *pET-Duet-1* expression plasmid to form *pETDuet-DgAS* (Figure 1a), which was then transformed into *E. coli* (DE3) with the electroporation method [20]. The recombinant *E. coli* (DE3) was cultured in Luria-Bertani (LB) medium to optical density at 560 nm (OD_{560}) of 0.6, and then induced with 0.2 mM of IPTG. After further cultivation at 18 °C for 20 h, the cells were centrifuged at 4500× *g* and 4 °C for 20 min. The cell pellet was washed, and spun down twice with 50 mM of phosphate buffer (PB) at pH 6.8, and then broken with sonication via a Branson S-450D Sonifier (Branson Ultrasonic Corp., Danbury, CT, USA). The sonication program was carried out for five cycles of 5 sec on and 30 sec off at 4 °C. The mixture was then centrifuged at 15,000× *g* and 4 °C for 20 min to remove the cell debris. The supernatant containing the produced DgAS fused with a His-tag in its N-terminal was applied in a Ni^{2+} affinity column (10 i.d. × 50 mM, Ni Sepharose 6 Fast Flow, GE Healthcare, Chicago, IL, USA). The His-tag fused DgAS was washed with PB with 25 mM of imidazole and eluted with PB containing 250 mM of imidazole. The elution was then concentrated and desalted through Macrosep 10 K centrifugal filters (Pall, Ann Arbor, MI, USA). The concentration of the purified DgAS was determined with the Bradford method [20], and analyzed with SDS-PAGE (Figure 1b). The purified DgAS was stored in a final concentration of 50% glycerol at –80 °C before use.

4.3. Biotransformation of 8-OHDe by the Purified DgAS Enzyme

Biotransformation was carried out in 1 mL of reaction mixture containing 0.1 mg/mL of 8-OHDe, 12.5 μg/mL of DgAS, 150 mM of sucrose, and 50 mM of Tris, pH 8.0. The reaction was performed at 40 °C for 30 min. After reaction, the reaction mixture was analyzed with UPLC. To optimize the major compound production, 1 mg/mL of 8-OHDe and 125 μg of DgAS were used as the substrate at different temperatures, pH levels, and sucrose concentrations. To optimize the pH, 50 mM of acetate buffer (pH 5 and pH 6), phosphate buffer (pH 7), and Tris buffer (pH 8 and pH 9) were used.

4.4. UPLC

UPLC was performed with an Acquity® UPLC system (Waters, Milford, MA, USA). The stationary phase was the Kinetex® C18 column (1.7 μm, 2.1 i.d. × 100 mM, Phenomenex Inc., Torrance, CA, USA), and the mobile phase was 1% acetic acid in water (A) and methanol (B). The linear gradient elution condition was 0 min with 36% B to 7 min with 81% B at a flow rate of 0.2 mL/min. The detection condition was set at 254 nm.

4.5. Purification and Identification of the Biotransformation Product

One hundred milliliters of the reaction mixture (1 mg/mL of 8-OHDe, 125 μg/mL of DgAS, 300 mM of sucrose, 50 mM of phosphate buffer, pH 7) was carried out at 40 °C for 30 min. At the end of the reaction, equal volume of methanol was added into the reaction mixture to stop the reaction. Two hundred milliliters of the reaction mixture with 50% of methanol was applied in a preparative YL9100 HPLC system (YoungLin, Gyeonggi-do, Korea). The stationary phase was the Inertsil ODS 3 column (10 mM, 20 i.d. × 250 mM, GL Sciences, Eindhoven, The Netherlands), and the mobile phase was the same as those in the UPLC system, but with a flow rate of 15 mL/min. The detection condition was 254 nm, and the sample volume was 10 mL for one injection. The product from each run was collected, concentrated under vacuum, and lyophilized with a freeze dryer. From the 100 mL of reaction, 90 mg of the major compound was purified. The chemical structure of the major compound was determined with mass and nmR spectral analysis. The mass spectral analysis was performed on a Finnigan LCQ Duo mass spectrometer (ThermoQuest Corp., San Jose, CA, USA) with ESI. ^{1}H- and

^{13}C-NMR, DEPT, HSQC, HMBC, COSY, and NOESY spectra were recorded on a Bruker AV-700 nmR spectrometer (Bruker Corp., Billerica, MA, USA) at ambient temperature. Standard pulse sequences and parameters were used for the nmR experiments, and all chemical shifts were reported in parts per million (ppm, δ).

4.6. Determination of Aqueous Solubility and Stability

Aqueous solubility and stability were conducted with the methods we used in a previous study [20]. For aqueous solubility assay, the tested compound was vortexed in distilled deionized H_2O for 1 h at 25 °C. The mixture was analyzed with UPLC. For stability assay, a stock of the tested compound (100 mg/mL in dimethyl sulfoxide) was diluted 100-fold to a concentration of 1 mg/mL in 50 mM of Tris buffer at pH 8.0. Then, samples were taken out for the UPLC analysis at the determined interval times.

4.7. Determination of Anti-Inflammatory Activity

The murine macrophage RAW 264.7 cell line was maintained in Dulbecco's Modified Eagle Medium (DMEM) supplemented with 10% fetal bovine serum (FBS), 100 μg/L streptomycin, and 100 IU/mL penicillin at 37 °C in a 5% CO_2 atmosphere. The RAW 264.7 cells were seeded at a density of 5×10^5 cells/well in 24-well plates, and incubated for 12 h at 37 °C and 5% CO_2. Different concentrations of the tested isoflavones were added. After 1-h treatment, the cells were stimulated with 100 ng/mL of LPS for 24 h. Culture supernatants (equal volumes) were mixed with Greiss reagent at room temperature for 10 min, and then the absorbance was measured at the wavelength 540 nm using a microplate reader (Sunrise, Tecan, Männedorf, Switzerland), as described previously [15]. This analysis was performed in tetraplicate. Relative inhibition of NO production was calculated with the equation: Relative inhibition (%) = [(OD_{570} with LPS only–OD_{570} with both LPS and isoflavones)/(OD_{570} with LPS only–OD_{570} without LPS or isoflavone)] × 100%. An IC_{50} value means a concentration of the drug that exhibited 50% of inhibition.

5. Conclusions

8-OHDe-7-α-glucoside is successfully produced from *O*-glucosylation of 8-OHDe with recombinant DgAS of *Deinococcus geothermalis* with a cheap and abundant renewable substrate, sucrose, as a sugar donor. The isoflavone glucoside is more soluble and stable than those of 8-OHDe in working buffers. The long half-life of 8-OHDe-7-α-glucoside maintains moderate anti-inflammatory activity, and could be used for industrial applications in the future.

Supplementary Materials: The following are available online at http://www.mdpi.com/1420-3049/24/12/2236/s1, Table S1. nmR spectroscopic data for compound (**1**) (in DMSO-d6; 700 MHz); Figure S1. The ^{1}H-NMR (700 MHz, DMSO-d6) spectrum of compound (**1**); Figure S2. The ^{13}C-NMR (176 MHz, DMSO-d6) spectrum of compound (**1**); Figure S3. The DEPT-90 and DEPT-135 (176 MHz, DMSO-d6) spectra of compound (**1**); Figure S4. The HSQC (700 MHz, DMSO-d6) spectrum of compound (**1**); Figure S5. The HMBC (700 MHz, DMSO-d6) spectrum of compound (**1**); Figure S6. The H-H COSY (700 MHz, DMSO-d6) spectrum of compound (**1**); Figure S7. The H-H NOESY (700 MHz, DMSO-d6) spectrum of compound (**1**); Figure S8. The key HMBC (blue arrows) and NOESY (pink arrow) correlations of compound (**1**).

Author Contributions: Conceptualization: T.-S.C.; data curation: T.-S.C. and S.-Y.Y.; methodology: T.-S.C., Y.-H.K. and C.-M.C.; project administration: T.-S.C.; writing—original draft: T.-S.C., T.-Y.W., and C.M.C.; writing—review and editing: T.-Y.W., J.-Y.W., and C.-M.C.

Funding: This research was financially supported by grants from the National Scientific Council of Taiwan (project no. MOST 107-2622-E-024-002-CC3).

Conflicts of Interest: The authors declare no conflict of interest.

References

1. Franke, A.A.; Custer, L.J.; Cerna, C.M.; Narala, K.K. Quantitation of phytoestrogens in legumes by HPLC. *J. Agric. Food Chem.* **1994**, *42*, 1905–1913. [CrossRef]

2. Chang, T.S. Isolation, bioactivity, and production of *ortho*-hydroxydaidzein and *ortho*-hydroxygenistein. *Int. J. Mol. Sci.* **2014**, *15*, 5699–5716. [CrossRef] [PubMed]
3. Komiyama, K.; Funayama, S.; Anraku, Y.; Mita, A.; Takahashi, Y.; Omura, S. Isolation of isoflavonoids possessing antioxidant activity from the fermentation broth of *Streptomyces* sp. *J. Antibiot.* **1989**, *42*, 1344–1349. [CrossRef]
4. Kiriakidis, S.; Högemeier, O.; Starcke, S.; Dombrowski, F.; Hahne, J.C.; Pepper, M.; Jha, H.C.; Wernert, N. Novel tempeh (fermented soybean) isoflavones inhibit in vivo angiogenesis in the chicken chorioallantoic membrane assay. *Br. J. Nutr.* **2005**, *93*, 317–323. [CrossRef] [PubMed]
5. Klus, K.; Barz, W. Formation of polyhydroxylated isoflavones from the soybean seed isoflavones daidzein and glycitein by bacteria isolated from tempe. *Arch. Microbiol.* **1995**, *164*, 428–434. [CrossRef] [PubMed]
6. Esaki, H.; Onozaki, H.; Morimitsu, Y. Potent antioxidative isoflavones isolated from soybean fermented with *Aspergillus saitoi*. *Biosci. Biotechnol. Biochem.* **1998**, *62*, 740–746. [CrossRef] [PubMed]
7. Esaki, H.; Kawakishi, S.; Morimitsu, Y.; Osawa, T. New potent antioxidative *o*-dihydroxyisoflavones in fermented Japanese soybean products. *Biosci. Biotechnol. Biochem.* **1999**, *63*, 1637–1639. [CrossRef]
8. Hirota, A.; Taki, S.; Kawaii, S.; Yano, M.; Abe, N. 1,1-Diphenyl-2-picrylhydrazyl radical-scavenging compounds from soybean miso and antiproliferative activity of isoflavones from soybean miso toward the cancer cell lines. *Biosci. Biotechnol. Biochem.* **2000**, *64*, 1038–1040. [CrossRef]
9. Lo, Y.L. A potential daidzein derivative enhances cytotoxicity of epirubicin on human colon adenocarcinoma Caco-2 cells. *Int. J. Mol. Sci.* **2012**, *14*, 158–176. [CrossRef]
10. Chang, T.S.; Ding, H.Y.; Tai, S.S.K.; Wu, C.Y. Tyrosinase inhibitors isolated from soygerm koji fermented with *Aspergillus oryzae* BCRC 32288. *Food Chem.* **2007**, *105*, 1430–1438. [CrossRef]
11. Chang, T.S. Two potent suicide substrates of mushroom tyrosinase: 7,8,4′-trihydroxyisoflavone and 5,7,8,4′-tetrahydroxyisoflavone. *J. Agric. Food Chem.* **2007**, *55*, 2010–2015. [CrossRef] [PubMed]
12. Goh, M.J.; Park, J.S.; Bae, J.H.; Kim, D.H.; Kim, H.K.; Na, Y.J. Effects of *ortho*-dihydroxyisoflavone derivatives from Korean fermented soybean paste on melanogenesis in B16 melanoma cells and human skin equivalents. *Phytother. Res.* **2012**, *26*, 1107–1112. [CrossRef] [PubMed]
13. Tai, S.S.K.; Lin, C.G.; Wu, M.H.; Chang, T.S. Evaluation of depigmenting activity by 8-hydroxydaidzein in mouse B16 melanoma cells and human volunteers. *Int. J. Mol. Sci.* **2009**, *10*, 4257–4266. [CrossRef] [PubMed]
14. Fujita, T.; Funako, T.; Hayashi, H. 8-Hydroxydaidzein, an aldose reductase inhibitor from okara fermented with *Aspergillus sp.* HK-388. *Biosci. Biotechnol. Biochem.* **2004**, *68*, 1588–1590. [CrossRef] [PubMed]
15. Wu, P.S.; Ding, H.Y.; Yen, J.H.; Chen, S.F.; Lee, K.H.; Wu, M.J. Anti-inflammatory activity of 8-hydroxydaidzein in LPS-stimulated BV2 microglial cells via activation of Nrf2-antioxidant and attenuation of Akt/NF-κB-inflammatory signaling pathways, as well as inhibition of COX-2 activity. *J. Agric. Food Chem.* **2018**, *66*, 5790–5801. [CrossRef] [PubMed]
16. Kim, E.; Kang, Y.G.; Kim, J.H.; Kim, Y.J.; Lee, T.R.; Lee, J.; Kim, D.; Cho, J.Y. The antioxidant and anti-inflammatory activities of 8-hydroxydaidzein (8-HD) in activated macrophage-like RAW 264.7 cells. *Int. J. Mol. Sci.* **2018**, *19*, 1828. [CrossRef] [PubMed]
17. Seo, M.H.; Kim, B.N.; Kim, K.R.; Lee, K.W.; Lee, C.H.; Oh, D.K. Production of 8-hydroxydaidzein from soybean extract by *Aspergillus oryzae* KACC 40247. *Biosci. Biotechnol. Biochem.* **2013**, *77*, 1245–1250. [CrossRef] [PubMed]
18. Wu, S.C.; Chang, C.W.; Lin, C.W.; Hsu, Y.C. Production of 8-hydroxydaidzein polyphenol using biotransformation by *Aspergillus oryzae*. *Food Sci. Technol. Res.* **2015**, *21*, 557–562. [CrossRef]
19. Chang, T.S. 8-Hydroxydaidzein is unstable in alkaline solutions. *J. Cosmet. Sci.* **2009**, *60*, 353–357. [CrossRef]
20. Chiang, C.M.; Wang, T.Y.; Yang, S.Y.; Wu, J.Y.; Chang, T.S. Production of new isoflavone glucosides from glycosylation of 8-hydroxydaidzein by glycosyltransferase from *Bacillus subtilis* ATCC 6633. *Catalysts* **2018**, *8*, 387. [CrossRef]
21. Das, S.; Rosazza, J.P. Microbial and enzymatic transformation of flavonoids. *J. Nat. Prod.* **2006**, *69*, 499–508. [CrossRef] [PubMed]
22. Cao, H.; Chen, X.; Jassbi, A.R.; Xiao, J. Microbial biotransformation of bioactive flavonoids. *Biotechnol. Adv.* **2015**, *33*, 214–223. [CrossRef] [PubMed]
23. Shimoda, K.; Hamada, H.; Hamada, H. Synthesis of xylooligosaccharides of daidzein and their anti-oxidant and anti-allergic activities. *Int. J. Mol. Sci.* **2011**, *12*, 5616–5625. [CrossRef] [PubMed]

24. Tiwari, P.; Sangwan, R.S.; Sangwan, N.S. Plant secondary metabolism linked glycosyltransferases: an update on expanding knowledge and scopes. *Biotechnol. Adv.* **2016**, *34*, 716–739. [CrossRef] [PubMed]
25. Hofer, B. Recent developments in the enzymatic *O*-glycosylation of flavonoids. *Appl. Microbiol. Biotechnol.* **2016**, *100*, 4269–4281. [CrossRef]
26. Claire, M.; Isabelle, A.; Magali, R.S. GH13 amylosucrases and GH70 branching sucrases, atypical enzymes in their respective families. *Cell. Mol. Life Sci.* **2016**, *73*, 2661–2679.
27. Stephane, E.; Sophie, M.; Kais, J.; Isabelle, A.; Pierre, M.; Magali, R.S.; Gabrielle, P.V. Cloning, purification and characterization of a thermostable amylosucrase from *Deinococcus geothermalis*. *FEMS Microbiol. Lett.* **2008**, *285*, 25–32.
28. Seo, D.H.; Jung, J.H.; Ha, S.J.; Cho, H.K.; Jung, D.H.; Kim, T.J.; Baek, N.I.; Yoo, S.H.; Park, C.S. High-yield enzymatic bioconversion of hydroquinone to α-arbutin, a powerful skin lightening agent, by amylosucrase. *Appl. Microbiol. Biotechnol.* **2012**, *94*, 1189–1197. [CrossRef]
29. Lee, H.S.; Kim, T.S.; Parajuli, P.; Pandey, R.P.; Sohng, J.K. Sustainable production of dihydroxybenzene glucosides using immobilized amylosucrase from *Deinococcus geothermalis*. *J. Microbiol. Biotechnol.* **2018**, *28*, 1447–1456.
30. Kim, M.D.; Jung, D.H.; Seo, D.H.; Jung, J.H.; Seo, E.J.; Baek, N.I.; Yoo, S.H.; Park, C.S. Acceptor specificity of amylosucrase from *Deinococcus radiopugnans* and its application for the synthesis of rutin derivatives. *J. Microbiol. Biotechnol.* **2016**, *26*, 1845–1854. [CrossRef]
31. Kim, E.R.; Rha, C.S.; Jung, Y.S.; Choi, J.M.; Kim, G.T.; Jung, D.H.; Kim, T.J.; Seo, D.H.; Kim, D.O. Enzymatic modification of daidzin using heterologous expressed amylosucrase in *Bacillus subtilis*. *Food Sci. Biotechnol.* **2018**, *28*, 165–174. [CrossRef] [PubMed]
32. Cho, H.K.; Kim, H.H.; Seo, D.H.; Jung, J.H.; Park, J.H.; Baek, N.I.; Kim, M.J.; Yoo, S.H.; Cha, J.; Kim, Y.R.; et al. Biosynthesis of catechin glycosides using recombinant amylosucrase from *Deinococcus geothermalis* DSM 11300. *Enz. Microbial Tech.* **2011**, *49*, 246–253. [CrossRef] [PubMed]
33. Kim, K.H.; Park, Y.D.; Park, H.; Moon, K.O.; Ha, K.T.; Baek, N.I.; Park, C.S.; Joo, M.; Cha, J. Synthesis and biological evaluation of a novel baicalein glycoside as an anti-inflammatory agent. *Eur. J. Pharm.* **2014**, *744*, 147–156. [CrossRef] [PubMed]
34. Hamalainen, M.; Nieminen, R.; Vuorela, P.; Heinonen, M.; Moilanen, E. Anti-inflammatory effects of flavonoids: genistein, kaempferol, quercetin, and daidzein inhibit STAT-1 and NF-κB activations, whereas flavone, isorhamnetin, naringenin, and pelargonidin inhibit only NF-κB activation along with their inhibitory effect on iNOS expression and NO production in activated macrophages. *Med. Infla.* **2007**. [CrossRef]

Sample Availability: 1 mg of 8-OHDe-7-α-glucoside for each request is available from the authors.

Review

Halogenating Enzymes for Active Agent Synthesis: First Steps Are Done and Many Have to Follow

Alexander Veljko Fejzagić [1], Jan Gebauer [2], Nikolai Huwa [1] and Thomas Classen [1,*]

1 Institute for Bio- and Geosciences I: Bioorganic Chemistry, Forschungszentrum Jülich GmbH, D-52426 Jülich, Germany; a.fejzagic@fz-juelich.de (A.V.F.); n.huwa@fz-juelich.de (N.H.)
2 Institute of Bioorganic Chemistry, Heinrich Heine University Düsseldorf, D-52426 Jülich, Germany; j.gebauer@fz-juelich.de
* Correspondence: t.classen@fz-juelich.de; Tel.: +49-2461-61-2208

Academic Editor: Josefina Aleu
Received: 11 October 2019; Accepted: 31 October 2019; Published: 5 November 2019

Abstract: Halogens can be very important for active agents as vital parts of their binding mode, on the one hand, but are on the other hand instrumental in the synthesis of most active agents. However, the primary halogenating compound is molecular chlorine which has two major drawbacks, high energy consumption and hazardous handling. Nature bypassed molecular halogens and evolved at least six halogenating enzymes: Three kind of haloperoxidases, flavin-dependent halogenases as well as α-ketoglutarate and *S*-adenosylmethionine (SAM)-dependent halogenases. This review shows what is known today on these enzymes in terms of biocatalytic usage. The reader may understand this review as a plea for the usage of halogenating enzymes for fine chemical syntheses, but there are many steps to take until halogenating enzymes are reliable, flexible, and sustainable catalysts for halogenation.

Keywords: bromination; chlorination; pharmaceuticals; active agent synthesis; biocatalysis; haloperoxidase; halogenase

---

1. Still Up-to-Date—Halogens in Active Agents

For the discovery of new active agents, synthetic chemists frequently look into natural compounds and deduce lead structures and functionalities for the assembly of active agent libraries. Although most natural compounds are not halogenated, halogenation is spread over virtually all classes of secondary metabolites. Most of the halogenated natural compounds are of marine origin, while some are found in plants and insects as well [1]. Halogens appear in some form in 40% of all drugs being tested in clinical trials [2–4]. In addition to the fact that halogenations are an important structural motifs in natural substances and thus also in the resulting active substances, halogenations play a major role in the synthesis of many active substances. In the following, we want to figure out what is so special about the simple halogen moieties within molecules and reactions that make them so desirable, although the synthesis is very energy-demanding and carried out with toxic molecular halogens such as chlorine gas. In the second half of this review article we would like to show how nature realizes halogenations enzymatically and where we stand technologically to employ them as tools. In recent years, these enzymes have become even more prominent and the various scientific advances in this field have already been presented several times in an overview. These reviews also provide an up-to-date overview of the different enzymes, their substrate scope and biotechnological developments as well as the diversity of halo-compounds from all kingdoms of life [5–14]. The aim of this review is—among other things—to include a further point of view. In addition to the accurate arguments on the toxicity of elemental halogens and the cost-effectiveness of halide salts, a closer look at the actual costs of chlorine gas production was included, as well as a clear presentation that chemical halogenating reagents are all based on the provision of these halogen gases. In addition, the most recent achievements for industrial applications e.g., by up-scaling processes, but also the distribution of these enzymes, as well as the break with assumed dogmas, such as conserved structural motifs, were taken into account.

1.1. Electronical Properties of Halogen Moieties

The presence of a halogen (Cl, Br, I) usually increases the bulkiness of a compound, blocking for instance active site pockets or increases membrane permeability, relevant for oral absorption, and blood–brain barrier permeability. Besides their bulkiness, halogens exhibit extraordinary effects on the polarization of a compound. On the one hand, the halogens of the upper periods (F, Cl, Br) have a large electronegativity, which leads to a considerable latent polarization in the molecule (see Figure 1A). On the other hand, the polarizability increases with increasing period, so that interactions with soft nucleophiles or electrophiles in particular are promoted (see Figure 1B). Although the latent polarization is depicted in Figure 1A as a homogeneous gradient the model must be refined. Due to the p-orbital architecture there is a hole in the electron density opposing the binding partner of the halogen which is called the σ-hole (Figure 1C). Considering this σ-hole, it offers the option to interact with heteroatoms (O, S, N) by so-called halogen bonds as well as hydrogen bonds [15]. The ability to form halogen bonds has been the focus of several pharmacologically-oriented groups in the past years, as it can serve as an alternative non-covalent interaction between atoms (see Figure 1). For a detailed insight into the nature and characteristics of halogen bonds, as well as their possible impact on drug discovery in the future, see the corresponding articles [4,15–19]. The importance of halogens for biological activity of compounds can be profound. Vancomycin (**1**, Figure 2), an antibiotic, was shown to exhibit 30% to 50% less activity, based on the chlorine substituents missing, which is remarkable considering how small the portion of the halogens with respect to the entire vancomycin molecule is [20].

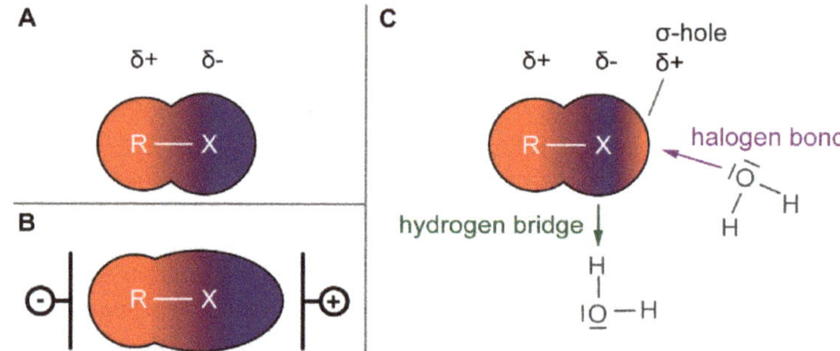

Figure 1. Schematic representation of electron distribution in halogens. (**A**): Latent polarization of a carbon-halogen bond. (**B**): Polarizability of large halogens (Br, I) bonded with a carbon. The external electrical field, for example, caused by an approaching electrophile/nucleophile leads to the distortion of the electron density. (**C**): Schematic view on the "σ-hole". The electron density is drawn to the carbon-halogen bond, with the strength gradually increasing with the size of the halogen (I > Br > Cl >> F). This anisotropic distribution of electrons in the outer orbitals of the halogen creates an area of higher electron density around the belt of the halogen, allowing interaction with electrophiles or H-bonds. Orthogonal to the direction of the bond is an area of electron deficiency, creating a partially positively charged area in the halogen, allowing for nucleophilic attacks, commonly called "σ-hole".

In terms of drug discovery, halogen substituents are regularly found in promising drug candidates with 35% in the discovery stage, while they appear in 36% of the candidates in clinical phase II and 26% in the drugs launched into the market (data from 2014) [16]. This trend shows that halogens play an important role in the field of drug design and discovery, and usually find their way to the final product assigned for treatment. In the following paragraphs, relevant halogens and some associated drug candidates containing halogen atoms will be discussed regarding their characteristic effects on bioactivity.

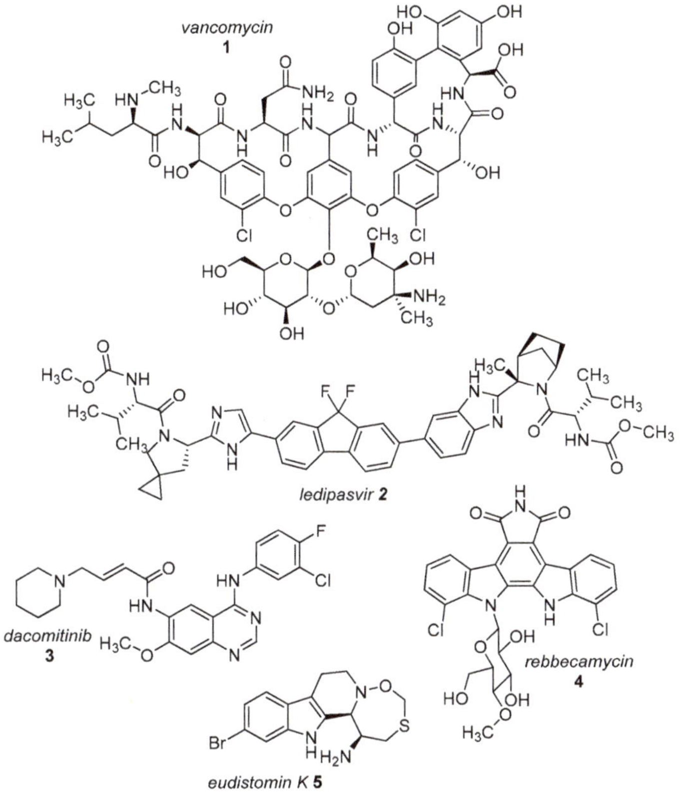

Figure 2. Examples for halogenated active agents.

The most prominent halogen introduced into active agents is fluorine with 57% [3]. Due to its similar size compared to hydrogen and the extreme electronegativity, C–F bonds are polarized in a distinctive manner and render fluorine a weak halogen bond acceptor in contrast to be a good hydrogen-bond acceptor [21,22]. The covalent fluorine bond is very strong (456 kJ/mol for CF_4), so that these bonds can only be cleaved under extreme and costly conditions in the body, if at all [23]. This increases the half time of active agents within the body (and environment) compared to their non-fluorinated pendants. Besides the electronic effects of fluorine within a molecule, fluorine also provides stereochemical properties which is summarized as fluorine *gauche* effect. Briefly, it can be described as a non-bonding weak interaction of the fluorine orbitals and other interacting partners. This reduces the degrees of freedom in rotation and this determines the conformation of a particular fluorinated molecule or guides reaction pathways. A review concerning this topic can be found in reference [24]. Apart from altering molecular characteristics, ^{18}F is used as a common radioactive isotope label for in vivo study of protein function and enzyme catalysis [25]. Of all halogenated active agents, ledipasvir (**2**, see Figure 2) is one of the top-selling drugs, administered for the treatment of hepatitis C. Another important compound is dacomitinib (**3**), a single-fluorinated drug, which has been in clinical trials for the treatment of non-small-cell lung cancer [26].

Chlorine is the second prevalent halogen with 38% in halogenated drugs. Due to its increased size, it is a moderate halogen bond acceptor, while still being stable when being introduced into a carbon bond (327 kJ/mol for CCl_4) [23]. Its presence in a compound alters volume and shape, allowing for positioning in deep cavities within proteins. These characteristics make it an interesting option for the functionalization of heterocycles. One of the most prominent chlorine-based natural compounds is rebeccamycin (**4**), a weak topoisomerase I inhibitor, which showed significant antitumor properties [27].

Brominated compounds are rarely found in drugs, making up only 4% of all halogenated compounds. This seems contradictory at first, as most halogenated compounds originate from marine organisms and are brominated despite chlorine being the more abundant halogen in water. Due to the lower polarization of the carbon-bromine bond and the extended bulkiness, bromine usually forms longer and thereby more labile bonds, not suitable for most drug candidates for a proper inhibition (272 kJ/mol for CBr_4) [23]. These characteristics however allow an easier oxidation of bromine and consequently an easier incorporation into molecules, compared to chlorine. Although there is a prevalence of chlorinated and fluorinated active agents in pharmacology, some brominated compounds are known to display relevant bioactivity like eudistomin K (**5**), viable for the treatment of polio and herpes [28].

Iodine is the rarest halogen used (1%), commonly exploited for the synthesis of the active agents. Having a higher size and lower electronegativity, its bonds formed with carbon atoms are more labile than those of bromine, being easily cleaved off. Iodine is, therefore, preferably suitable for short-lived applications. An example of the use of iodine in medicine is radioactively labelled ^{124}I in positron emission tomography (PET) as a tracer [29].

1.2. Halogens as Synthetic Tools

Both, bromine and iodine, are rare as functional moieties in active agents due to their labile covalent bonds. But it is precisely these properties that make halogens of higher periods valuable instruments for the synthesis of active substances.

A patent application for the production of hypohalous acids was applied for in 1944. *C. C. Crawford* and *T. W. Evans* described a process to obtain halide-free solutions of hypochlorous acid. This halogenating reagents were used in industrial applications to produce e.g., halohydrins from unsaturated organic compounds [30]. In 1993 another patent to produce concentrated slurries of sodium hypochlorite [35% (*w*/*v*)] was accepted [31]. They describe a process for highly pure hypochlorite slurry production. All the processes have the same starting materials in common. The first step is the solvation of molecular chlorine in water to get hypohalous acid (**6**) or the solution of sodium hydroxide and chlorine in water to end up with sodium hypochlorite. However, contaminations of sodium chloride and remaining sodium hydroxide occur in most processes that are carried out in industrial scale. The chlorine is hereby acquired by the chloralkali process where the electrolysis of sodium chloride produces molecular chlorine gas. Similar processes are state-of-the-art for the production of sodium bromate, which has the drawback of being a strong oxidizing agent [32,33], but can be used for the bromination of aromatic compounds [34]. The production of stable hypobromous acid is rather difficult because it easily oxidizes to bromate. Here, the production is carried out starting from hypochlorous acid or a modified chlorite [35].

More common halogenating agents are *N*-bromo-succinimide (NBS) and *N*-chloro-succinimide (NCS). Interestingly, even these reagents are synthesized from molecular halogens or hypohalous acids [36]. As a conclusion, it is now rather obvious, that all halogenating reagents have their origin in molecular halogen gases that are produced by cost-intensive procedures like halogen alkali electrolysis from halide salts (Figure 3).

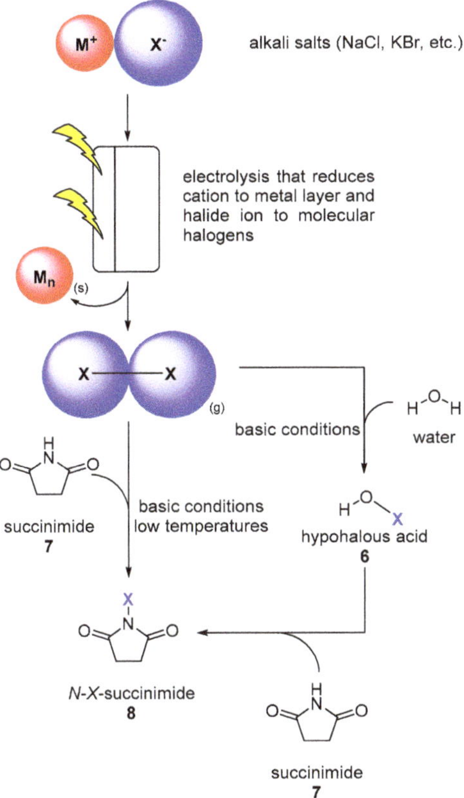

Figure 3. Workflow for the provision of halogenating reagents from alkali salts. The electrolysis process thus produces molecular halogens (X_2), as well as hypohalous acids (HOX, **6**) and *N*-halogenated succinimides (NXS, **8**) in further steps.

Having these halogenated building blocks at hand, further synthetic steps can follow to build up active agents. Not only in academia but also in industry, the synthetic tool in terms of cross-coupling reactions is one of the most common C–C- and C–Y bond formations (Y is in this case N, O, S). With the use of different transition metals and activated carbon components, it is possible to generate large bioactive natural products and their derivatives. One prominent example is the use of palladium for the selective preparation of arenes and heterocyclic scaffolds with different substitution patterns [37]. However, also non-noble transition metals like copper [38], nickel [39], and nowadays even iron [40–42] are firmly anchored as suitable catalysts. Besides the high chemoselectivity, a profound functional group tolerance is a main advantage of these kind of reactions. Therefore, it is not surprising that industry has established approaches to produce pharmaceuticals and fine chemicals at the kilogram scale [43,44]. The following Figure 4 gives an overview of the most popular metal catalyzed named reactions, that slightly differ in their reactive moieties for both products or starting materials [37,45–48]. However, the catalytic cycle and thereby the reaction mechanism is very similar for all (Figure 5). Finally, conversions such as the *Appel* reaction and the *Hell-Volhard-Zelinsky* reaction, in which functional groups such as alcohols are converted to haloalkanes or carboxylic acids that become acid chlorides, must also be mentioned here.

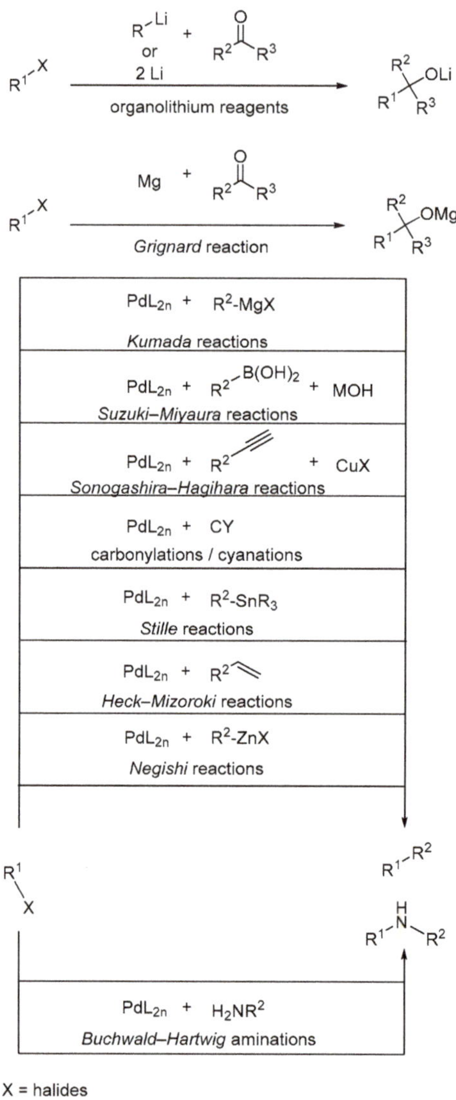

Figure 4. Most common reactions in organic synthesis exploiting halogen moieties. [37,45–48] Besides organolithium reactions as well as Grignard/Barbier reactions all of them are Pd-based, but can in many cases be substituted by other transition metals such as nickel.

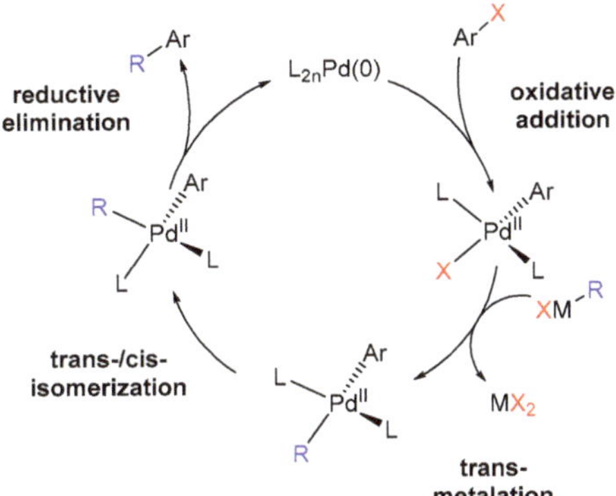

Figure 5. Scheme of the steps in cross-coupling reactions. After oxidative addition of the organo-halogen species, the transmetalation occurs. The ligands start rearrange before reductive elimination to the final product is carried out and the catalyst is regenerated.

1.3. Halogen Chemistry is Energy-Demanding

It is estimated that about 55% of chemical and 85% of pharmaceutical end products were processed with key components derived from the chloralkali electrolysis process [49,50]. These include hydrochloric acid to adjust the pH, or chlorinated solvents as part of the synthesis and subsequent isolation. However, this results in the production of the active compounds under hazardous conditions and high costs, due to toxic waste management. Using enzymes to halogenate pharmaceutical active compounds in a mild way and with a high efficiency is certainly a desirable aim for a greener chemistry. In general, the production and further processing of chlorine is mostly performed in the very same geographical region or facility in order to avoid the transportation of toxic and dangerous intermediates. This was reported for German companies and, presumably, this is also the case for other countries. The key component for halogenation (chlorine) is produced by electrolysis and is one of the most energy-consuming processes in the chemical industry. The process is responsible for about 2% of the total energy consumption yielding 5 million tons per year of chlorine in Germany [50,51]. Obviously, the energy reduction is an objective of the chloralkali industry, since 50% to 60% of the production costs is spend for the electrical energy [52].

2. Halogenating Enzymes

Although halogenated natural compounds are rare and only found within the regime of secondary metabolism, at least six types of halogenating enzymes were evolved. Many were evolved from monooxygenases, since hypohalous acids are the core intermediate of catalysis in these halogenating enzymes. As diverse the origins of halogenating enzymes are as diverse is their classification. In Figure 6 we tried to give an overview on the categories of halogenating enzymes. Although often used synonymously, it can be differentiated between haloperoxidases and halogenases. The first group forms hypohalous acid from the respective halide and hydrogen peroxide via heme-iron-, vanadium-coenzymes, or even without any coenzyme. The hypohalous acid is set free for most of the enzymes and the very halogenation reaction takes place outside the active site. In contrast, the halogenases generate or simply use halonium species for the halogenation without the use of hydrogen peroxide.

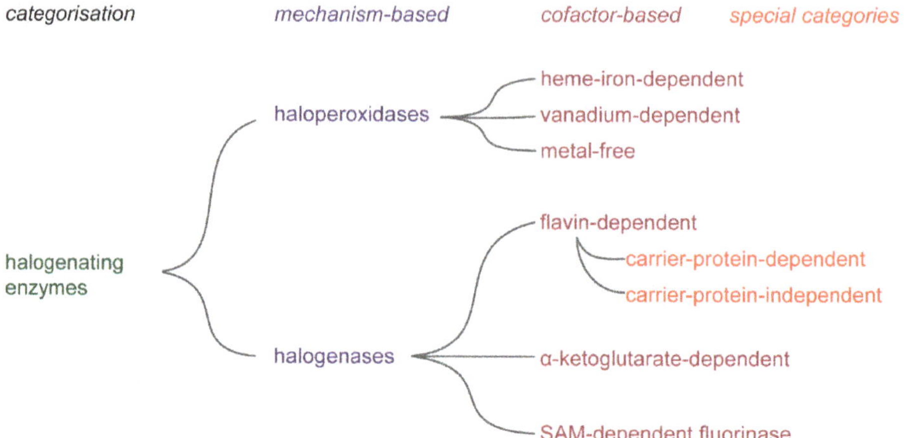

Figure 6. Overview on the categorization of halogenating enzymes.

Haloperoxidases

Haloperoxidases were the first group of halogenating enzymes discovered in the past. Enzymes of this family catalyze the oxidation of a halide anion (X^-) in presence of hydrogen peroxide to an oxidized halide form, usually believed to be the corresponding hypohalous acid. The class is further subdivided into three subclasses, the heme-iron-dependent, vanadium-dependent, and metal-free haloperoxidases or perhydrolases. In the following section, each class will be discussed briefly with biocatalytic examples, if they are known.

3. Heme-Iron-Dependent Haloperoxidases

The heme-iron-dependent haloperoxidases were the first and most intensively studied haloperoxidases. Back in the 1960s, an enzyme from the fungus *Caldariomyces fumago* (*Leptomyxes fumago*) was shown to be responsible for the halogenation of 1,3-cyclopentadion to the natural compound caldariomycin (**9**) [53]. Upon further investigation, it could be shown that it contained a heme-prosthetic group tethered to the enzyme by a distal cysteine ligand, very similar to the P450 monooxygenases [54].

Cl Cl
HO OH
caldariomycin
9

The catalytic cycle (Figure 7) displays a key intermediate, the Fe^{IV}-oxo-species, to oxidize chlorine to hypochlorite, which is released and may be attacked by an electron-rich substrate serving as an electrophile. In presence of excess hydrogen peroxide, this complex can alternatively decompose to molecular oxygen and chloride.

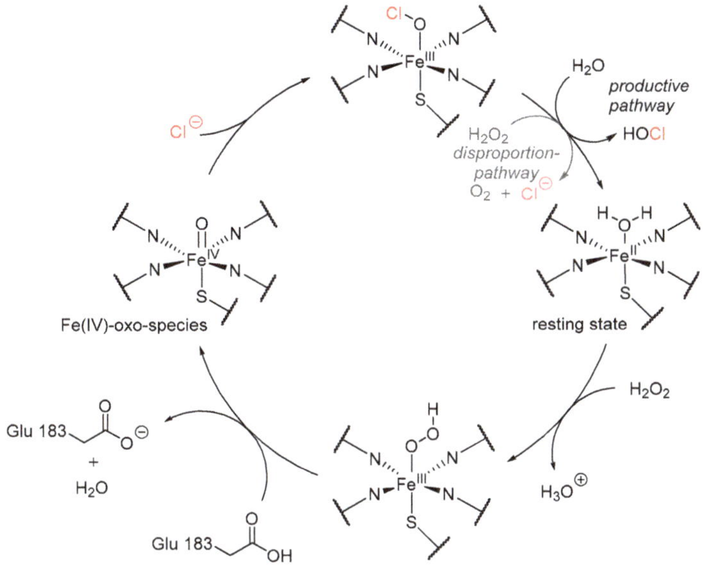

Figure 7. Proposed catalytic cycle of heme-iron-dependent haloperoxidases, shown on the example of CPO from *C. fumago*. In the resting state (3 o' clock), water is bound to the heme-iron, which is subsequently replaced by hydrogen peroxide. After protonation of this complex by a catalytic glutamate (Glu183), water is eliminated, creating the actual active species, the Fe(IV)-oxo complex. A halide, in this case chloride, binds to the Fe(IV)-oxo species and is released as hypochloric acid, regenerating the heme-site by hydrolysis with water. Alternatively, another molecule hydrogen peroxide may attack, leading to the disproportion of the complex to molecular oxygen, water, and chloride [54,55].

As the enzyme resembled characteristics from peroxidases as well as monooxygenases, it was classified as a heme-iron-dependent haloperoxidase and due to its ability to oxidize all halides besides fluorine was named chloroperoxidase. Recently it was revealed that actually two *Cf-cpo* genes within the *C. fumago* genome exist, sharing a high sequence identity and both being present in the secreted supernatant of its host [56]. Since its discovery, the enzyme was target of many mechanistic and biocatalytic studies. To much surprise, the formed hypohalous acid does not leave the active site freely, but is held back by amino acids placed in the halide entrance tunnel of *Cf*-CPO, allowing for regio- and enantioselectivity to a certain degree, mainly depending on the nature of the substrate [57]. Its major drawback, however, was the oxidative inactivation every heme-iron-containing protein suffers after exposure to oxygen as well as a high sensitivity for high hydrogen peroxide concentrations. As the genetic modification of the fungus can prove tedious, the application of this enzyme in biocatalysis might seem limited, however due to the fruitful work of Pickard et al., protocols are available for a reasonable production and secretion of the enzyme in the native host, *C. fumago* [58].

As a catalyst, *Cf*-CPO was shown to be rather robust und allow a variety of different organic transformations, where some are not always bound to a halogenating step. It could be applied in cascade reactions with oxidases leading to halocyclization reactions of allenes (**10**) and even be immobilized for (semi-)continuous-flow bioreactors [59–61] (see Figure 8). It was used for the halogenation of phenolic monoterpenes like thymol (**12**) and carvacrol, excelling with drastically lower catalyst loadings (by five orders of magnitude) compared to chemical alternatives like Cu^{II}-catalysis [59]. Furthermore, it was shown to be capable of halogenating *trans*-cinnamic acid and other unsaturated carboxylic acids, as well as catalyze enantioselective epoxidation of alkenes [62,63]. One bottleneck observed was the low substrate loading, impairing possible preparative applicability.

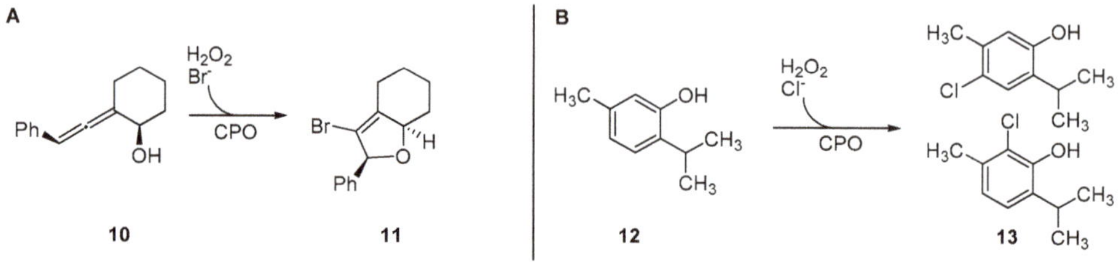

Figure 8. Example reactions of *Cf*-CPO involved in biocatalytic conversions of organic molecules (**A**): Cyclization of allenes (**10**) to the product **11** induced by halogenation with Br^- by *Cf*-CPO. (**B**): Unselective chlorination of thymol (**12**) by *Cf*-CPO.

Besides chloroperoxidase from *C. fumago*, not many members of this subclass have been dealt with. The bromoperoxidases from *Pseudomonas aureofaciens* and *Penicillus capitatus* are other examples of such heme-iron-dependent enzymes [64,65]. However, beside classic characterization experiments, revealing similar properties to *Cf*-CPO such as high thermal stability and sensitivity to high hydrogen peroxide concentrations, no complex biotransformations were investigated with these enzymes, yet (see Table 1) [66].

Table 1. Enzymological properties of heme-dependent haloperoxidases (* original host).

Enzyme	Expression		Kinetic Parameters	Substrates
	Host	Yield [mg/L]		
Cf-CPO	*C. fumago* *	430 [67]	0.78 mM h^{-1} [59]	aromatic, alkenes
	E. coli BL21(DE3)	n.a. [68]		
	Aspergillus niger	10 [68]		
BPO	*Pseudomonas aureofaciens* [65]	n.a.	n.a. partial diastereo-selectivity [69]	aromatic, *N*-hetero-cycles, alkenes
	Penicillus capitatus [64,65]	n.a.		

4. Vanadium-Dependent Haloperoxidases

For several years after the discovery of heme-iron-dependent haloperoxidases, it was assumed that they are the only enzymes able to oxidize halides for the subsequent halogenation reaction. However, a new halogenating enzyme class was discovered in 1993 by van Schijndel et al. from *Curvularia inaequalis* using *ortho*-vanadate cofactor for the oxidation of halides [70,71]. Just two years later, a vanadate-dependent homolog from *Corallina officinalis* was crystallized [72]. These vanadium-dependent haloperoxidases became a popular research target as they were shown to exhibit high turnover numbers without suffering an oxidative inactivation and displaying a higher tolerance against hydrogen peroxide [73], In contrast to the heme-iron-dependent ones, however, they usually do not retain the formed hypohalous acid within the active site, leading to a freely diffusible strong oxidant. Resulting from this mechanistic aspect, random halogenations occur, even in the protein itself, leading to its destabilization and inactivation. Because of this free hypohalous species, the selectivity of the subsequent halogenation reaction is independent of the enzyme but from the electronic properties of the substrate. Most of the vanadium-dependent haloperoxidases originate from marine fungi and marine macroalgae (seaweeds) [74].

It is proposed that the catalytic cycle (Figure 9) forms a V^V-peroxo-species as the key intermediate, where the halide is added and subsequently hydrolyzed to hypohalous acid. Identically to heme-iron-dependent haloperoxidases, the presence of hydrogen peroxide may lead to the disproportion to singlet oxygen and the halide [55].

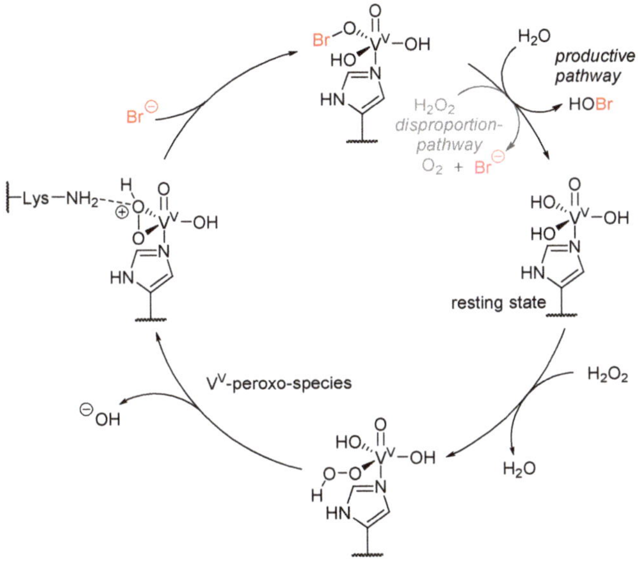

Figure 9. Proposed catalytic cycle of vanadium-dependent haloperoxidases. In its resting state (3 'o clock), vanadium contains four oxygen ligands, while the free coordination site is occupied by a catalytic histidine residue, resulting in a dative bond. In presence of hydrogen peroxide, a hydroxyl group is substituted by peroxide. Upon elimination of a hydroxide ion, a cycloperoxo-species is generated, which is stabilized by a catalytic lysine residue. This cyclic intermediate is opened by addition of a halide, in this case bromide, which can then be hydrolyzed by water, leading to the release of hypobromic acid, or in presence of another hydrogen peroxide molecule, be disproportioned to molecular oxygen and bromide. During catalysis, the vanadium does not alter its oxidation state (V) [55].

One of the best-investigated representatives of this class is the vanadium-dependent chloroperoxidase from the phytopathogenic fungus *Curvularia inaequalis* [70,71,75–77]. Even in absence of the vanadium-cofactor, the enzyme is stable in its *apo*-form and can easily be transformed to the holo-form by external addition of *ortho*-vanadate [70]. Although the gene can be heterologously expressed in *E. coli* and activated with vanadate, it was reported that the amount of enzyme obtained was very low. As an alternative, *Saccharomyces cerevisiae* was used as a host, yielding 100 mg/L *apo*-enzyme [75]. Kinetic experiments lead to a k_{cat}/K_M of 2.6×10^6 M^{-1} s^{-1} for hydrogen peroxide and 5.1×10^7 M^{-1} s^{-1} for bromide at pH 4.2, the optimal pH for bromoperoxidase activity [75].

It showed stability at high temperatures (T_M of 90 °C) and tolerance against organic solvents like methanol, ethanol, and propan-2-ol (up to 40% *v/v*) [71]. *Ci*-V_{Cl}PO was used as a hypohalogenite catalyst for the halogenation of phenols like thymol, while showing excellent stability towards hydrogen peroxide and organic solvents like methanol and ethyl acetate [76]. Furthermore, it was used for the mediation of *(Aza-)Achmatowicz* reactions in combination with cascades [78] and halofunctionalization reactions of aromatic and aliphatic alkenes like styrene and hexanol [77,79] (see Figure 10).

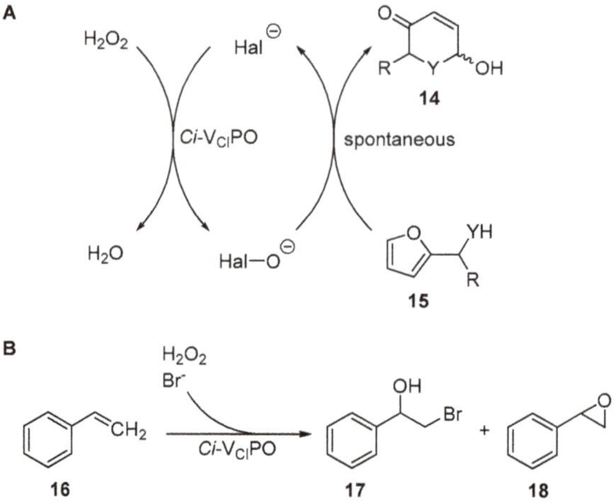

A

B

Figure 10. (**A**): *(Aza-)Achmatowicz* reaction transforming the furan **15** to the Michael-system [78] **14**. (**B**): Halofunctionalization of styrene (**16**) [77,79].

In contrast to the usually scarce selection of vanadium-dependent chloroperoxidases, many representatives of bromoperoxidases were researched in the past. One of the most prominent members of this group is the V_{Br}PO from *Corralina officinalis*, a marine red algae. Similarly to the homolog from *C. inaequalis*, it excels with a high stability towards high temperatures up to 90 °C and in presence of organic solvents like ethanol, propanol, and acetone (up to 40% *v/v*) [72]. However, recombinant expression of the gene in *E. coli* BL21(DE3) proved difficult, as the amount of protein formed is high, but insoluble. Coupe et al. notably showed that by using a refolding procedure, 40 mg/L of active enzyme can be retrieved after expression and isolation [80].

The haloperoxidase was shown to accept a variety of substrates, like nitrogen-containing heterocycles, cyclic β-diketones, phenol, *o*-hydroxybenzyl alcohols, anisole (**19**), 1-methoxynaphthalene and thiophene in addition to alkene halogenations with styrene (**16**), cyclohexene (**22**) among others to yield various bromohydrins [81] (see Figure 11 and Table 2).

In most of the cases, no diastereoselectivity for the bromohydrin formation could be observed, except for the formation of bromohydrin from (*E*)-4-phenyl-buten-2-ol (**24**) [69]. Besides bromination reactions, haloperoxidases like the *Co*-V_{Br}PO are able to catalyze sulfoxidations with 2,3-dihydrobenzothiopene (**26**), as well [82].

Table 2. Enzymological properties of vanadium-dependent haloperoxidases.

Enzyme	Expression		Kinetic Parameters	Substrates
	Host	Yield [mg/L]		
Ci-V_{Cl}PO	*C. inaequalis*	10 [70]	5.1×10^7 M^{-1} s^{-1} for Br^- [75]	aromatic, alkenes
	E. coli BL21(DE3)	15 [76]		
	S. cerevisiae	100 [75]		
Co-V_{Br}PO	*C. officinalis*		200 U/mg for MCD [83]	aromatic, *N*-hetero-cycles, alkenes
	E. coli BL21(DE3) (insoluble)	40		

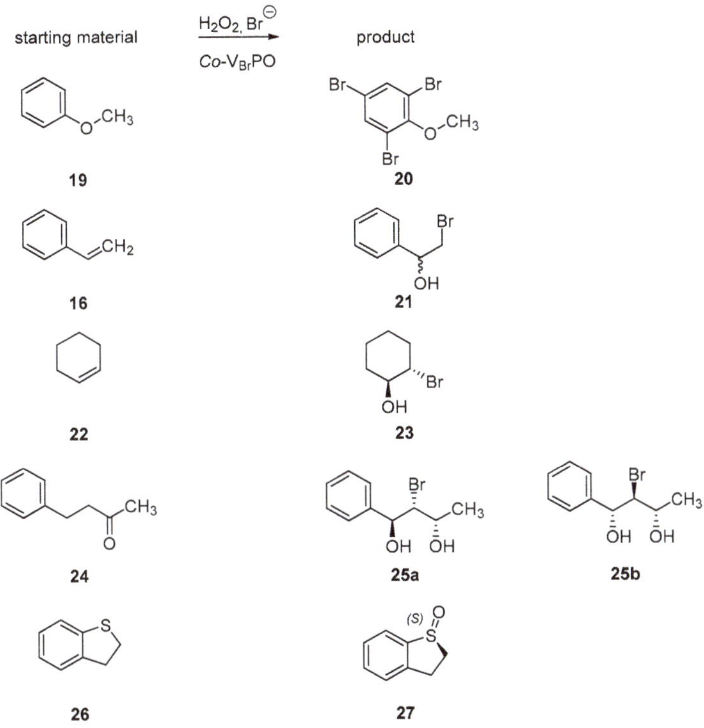

Figure 11. Selected reactions performed by the *Co*-V_{Br}PO to illustrate the reaction spectrum [81].

5. Metal-Free Haloperoxidases/Perhydrolases

Although oxidative halogenation reactions are dominated by (transition) metal catalysis in nature, a group of enzymes was identified catalyzing halogenation without any metal cofactor. These metal-free haloperoxidases or perhydrolases were found to require hydrogen peroxide and halides as well, while forming percarboxylic acids from carboxylic acids using a catalytic triad of serine, histidine, and aspartate [84,85]. Their striking resemblance to lipases has initiated a general debate over the nature of these enzymes, as their characteristics resemble hydrolases with a halogenating sub-activity. This has led to controversies whether the metal-free haloperoxidases are not simply lipase-like enzymes moonshining as haloperoxidases. In fact, several lipases were tested positively for haloperoxidase activity despite low turnover numbers [81].

The key-step in catalysis is the formation of a peroxo-acid from a carboxylic acid by hydrogen peroxide, which subsequently forms an acylhypohalide acting as the halogenating agent (Figure 12) [86].

Many examples for metal-free HPOs in biotransformations are not known. The majority of investigations of this enzyme class were focused on determining and expanding the tolerance of these enzymes to organic solvents and temperatures. One recent example of a bioorganic application was the halogenation of nucleobases and analogues [87] (see Figure 13).

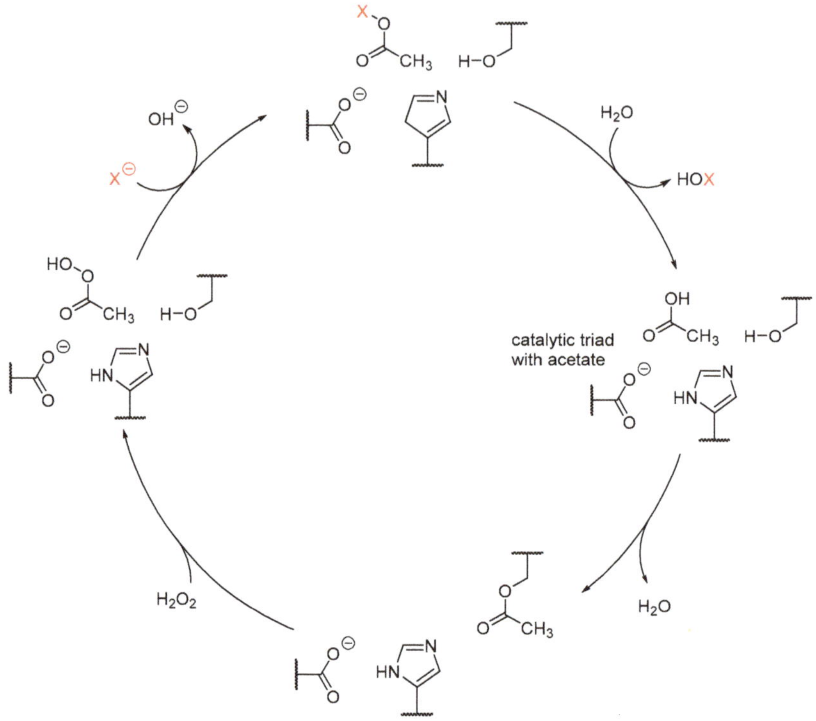

Figure 12. Proposed catalytic cycle of metal-free haloperoxidases/perhydrolases. This mechanism was compiled from several sources [81,85]. The catalytic cycle is adopted from the common hydrolase catalysis encountered in lipases and esterases, for instance. In presence of a carboxylic acid, in this case acetic acid, an ester is formed with the catalytic serin residue upon elimination of water (3 'o clock). In presence of hydrogen peroxide, the ester is cleaved, forming a percarboxylic acid. In the following step, a halide binds to the peroxoacid, which is hydrolyzed to the hypohalous acid, while the characteristic Ser-His-Asp triad is already regenerated.

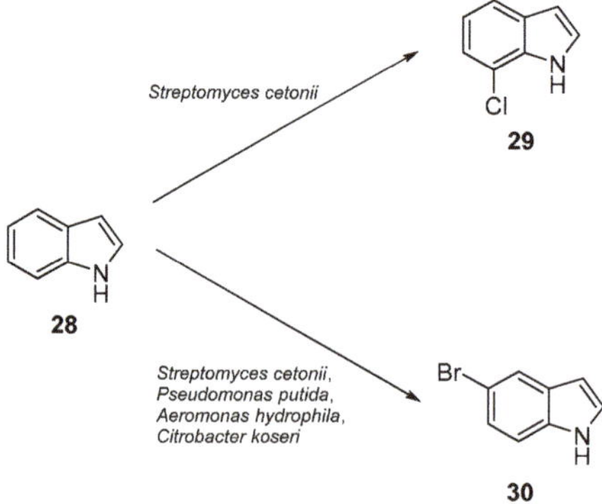

Figure 13. Halogenation of indole (**28**) nucleobase analogs according to Lewkowicz and co-workers [87].

5.1. Flavin-Dependent Halogenases

In addition to the long-known haloperoxidases, another class of enzymes has aroused much interest. It is suspected that flavin-dependent halogenases (FHals, Fl-Hals, or FDHs) evolved from monooxygenases that require flavin cofactors as well and, therefore, belong to the superfamily of flavin-dependent monooxygenases [88,89].

According to what is known so far, there are three natural target structures that can be addressed by FHals. The most studied and best understood group are the flavin-dependent tryptophan halogenases. In nature, there is the possibility to halogenate every position of the indole ring. Similar to this structure there is the group of flavin-dependent pyrrole halogenases and finally the flavin-dependent phenol halogenases (see Figure 14) [8]. The fact, that each and every position can be addressed by an individual enzyme demonstrates that FHals are selective halogenating catalysts in contrast to the majority of haloperoxidases. FHals must be differentiated according to the accessibility of their substrates. While a large number of these halogenases are involved in biosynthesis clusters of polyketides (PKS) and non-ribosomal protein synthesis (NRPS), some, such as tryptophan halogenases, can convert freely diffusible substrates and are not dependent on carrier proteins that activate or merely tether the substrate (Figure 14) [13].

For the application of this enzyme group, it is important to keep in mind that they need at least a two-component electron transport chain and therefore require a suitable flavin reductase [90–92]. In addition to the reductases that naturally belong to the biosynthesis clusters e.g., PrnF [93], applications with other reductases such as SsuE [91,94,95] or Fre [96,97] from *E. coli* have also been reported. To avoid the necessity of a second enzyme—the flavin reductase—or even a third enzyme for cofactor recycling, photochemical approaches are in the focus of current research in this field as well [98].

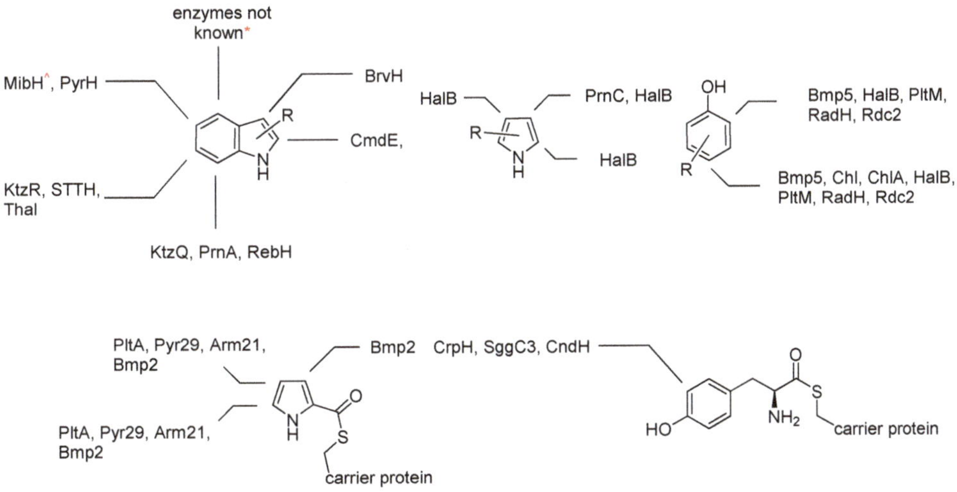

Figure 14. Regioselectivity of flavin-dependent halogenases and their dependency on carrier proteins. * Natural products with halogenations are known, but so far, no enzyme is characterized. ^ This tryptophan halogenase is one of the few examples that is carrier protein-dependent [99].

Figure 14 shows some representatives for the halogenation of the different positions of the different substrates (indoles [92,95,100–102], pyrroles [103] and phenols [104–107]), each with reference to the halogenating enzyme, the dependence on carrier proteins and the corresponding publication [108]. The halogenation of position four of indoles, as for example in 4-chloroindole-3-acetic acid, is known to date only from plants (*Pisum sativum*, *Lens culinaris*, *Vicia* sp., and in particular *Vicia faba*), as a growth hormone but no enzyme has yet been characterized responsible for its formation [7]. The publications e.g., by Shepherd et al., the review of Latham et al. and other publications also show various mutants that led to changes in regioselectivity and substrate scope [8,96,100,109,110].

A lot of these enzymes that are dependent on carrier proteins produce well-known secondary metabolites like rebeccamycin (**4**) and vancomycin (**1**) but also a plethora of less investigated biosynthetic pathways [133]. The most important difference in the mechanism between flavin-dependent monooxygenases and halogenases is the conserved motif of two tryptophanes, one isoleucine and proline. This 10 Å long tunnel [89], first found in PrnA, serves to spatially separate the activated peroxy-flavin FAD(C4α)–OOH from the substrate binding site and thus prevents oxygenation [114,116,128]. After generating the hypohalides, a conserved lysine transfers the electrophilic chlorine as chloramine from the former peroxy-flavin to the substrate (Figure 15) [134].

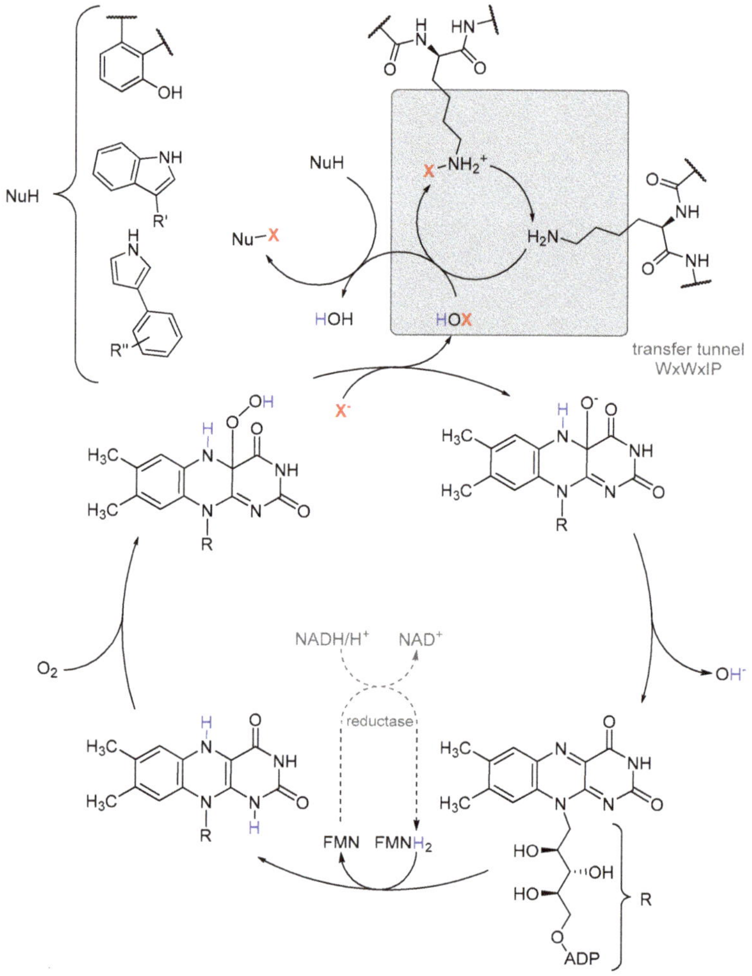

Figure 15. Catalytic cycle of halogenation by flavin-dependent halogenases [96,115].

For the phenol halogenase the mechanism is proposed to be slightly altered. The phenolic hydroxyl group is deprotonated by an aspartate within the active site increasing the nucleophilicity of the enol α-carbon [130]. Based on these conserved motifs and the assumed reaction mechanism some putative halogenases have already been found and annotated. Recently even a viral halogenase VirX1 from cyanophages was discovered, which is the first FHal capable of in vitro iodination and stands out due to its broad substrate spectrum and preferred iodination [135].

Although the community has so far agreed that the preserved motif of separation tunnel and anchor lysine seems to be essential for the activity, current research shows a further class of flavin-dependent halogenases which lack these structural elements completely. One of these examples is the halogenase KerK that is under investigation by Piel and coworkers but has not yet been published except as a poster presentation on Biotrans 2019 in Groningen, the Netherlands [136].

In addition to the advantages of high regioselectivity and thus only few by-products, there are also some disadvantages in the use of this enzyme group. The low conversion rates speak against large-scale application and expression problems often occur. Many of the proteins produced in *E. coli* BL21(DE3) end up in the insoluble fraction as inclusion bodies. To deal with this issue, strains with co-expression of chaperones are used regularly (Table 3). The overall stability of these proteins also needs further optimization to be applicable in biocatalysis. As a promising result Kemker et al. the tryptophan halogenases were successfully scaled up in terms of a biocatalytic process employing immobilizing the enzymes by cross-linked enzyme aggregates (CLEAs). This yielded L-7-bromotryptophan on the gram scale [137].

Table 3. Examples of flavin-dependent tryptophan, pyrrole and phenol halogenases that can be carrier-dependent or independent.

Enzyme	Origin	Heterologous Expression Host [1]	Product	Miscellaneous
PrnA [96]	*Pseudomonas fluorescens* BL915	*E. coli* ArcticExpress (DE3)	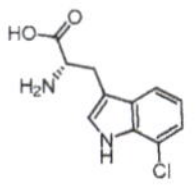	
RebH [92]	*Lechevalieria aerocolonigenes* strain 39243	*E. coli* BL21(DE3)	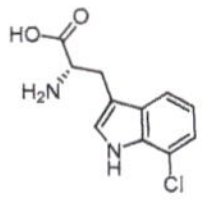	
KtzQ [110]	*Kutzneria* sp. 744		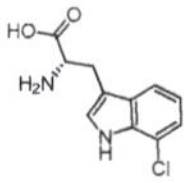	
KtzR [110]	*Kutzneria* sp. 744		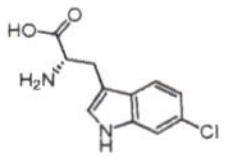	
CmdE [111]	*Chondromyces crocatus* Cm c5		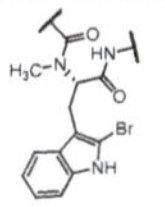	Post-NRPS (non-ribosomal peptides)
SSTH [101]	*Streptomyces toxytricini* NRRL 15443	*E. coli* BL21 CodonPlus (DE3)-RIL	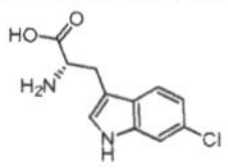	
Thal [112]	*Streptomyces albogriseolus*	*P. fluorescens* BL915 DORF1 and *P. chlororaphis* ACN	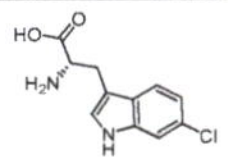	

Table 3. *Cont.*

Enzyme	Origin	Heterologous Expression Host [1]	Product	Miscellaneous
MibH [99,113]	*Microbispora coralline* NRRL 30420		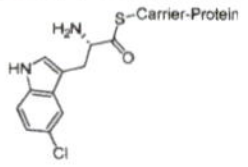	NRPS-dependent
PyrH [96,114]	*Streptomyces rugosporus* LL-42D005	*E. coli* ArcticExpress (DE3); *Pseudomonas fluorescens* BL915 ΔORF1	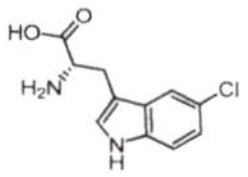	
Xcc-B B100XXXX [115]	*Xanthomonas campestris pv. campestris* strain B100	*E. coli* BL21(DE3) with pGro7 plasmid (Takara) for chaperone co-expression	Various substituted indoles and thereby differing regio-selectivity	
BrvH [116]	*Brevundimonas* BAL3	*E. coli* BL21(DE3) with pGro7 plasmid (Takara) for chaperone co-expression	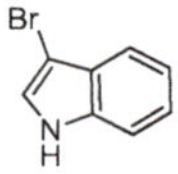	
PrnC [93,117]	*Pseudomonas fluorescens* BL915		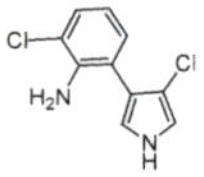	
Chl [118]	*Streptomyces aureofaciens*		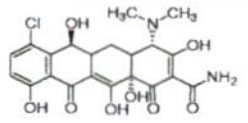	
ChlA [105]	*Dictyostelium discoideum*		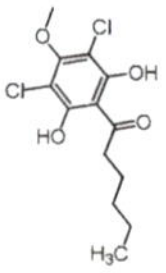	
PltA [119]	*Pseudomonas fluorescens* Pf-5	*E. coli* BL21(DE3)	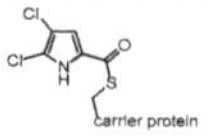	
Pyr29 [120]	*Actinosporangium vitaminophilum* ATCC 31673 and *Streptomyces* sp. Strain UC 11065		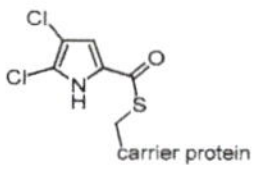	
Arm21 [121]	*Streptomyces armeniacus*		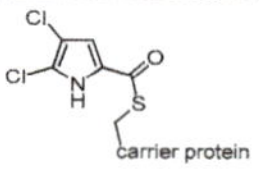	
CrpH [104]	*Nostoc* Cyanobionts		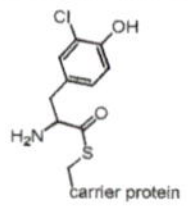	NRPS-dependent
BhaA [122,123]	*Amycolatopsis mediterranei* DSM5908		balhimycin	
SgcC3 [124]	*Streptomyces globisporus*	*E. coli* BL21(DE3) pET-30Xa/LIC	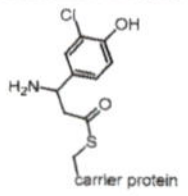	

Table 3. *Cont.*

Enzyme	Origin	Heterologous Expression Host [1]	Product	Miscellaneous
HalB [125]	*Actinoplanes* sp. ATCC 33002	*Pseudomonas aureofaciens* ACN	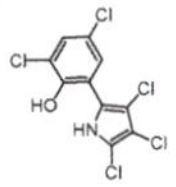	
PltM [126]	*Pseudomonas fluorescens* Pf-5	*E. coli* BL21 (DE3)	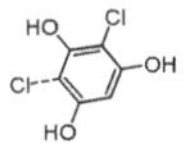	
Bmp5 [106]	*Pseudoalteromonas luteoviolacea*		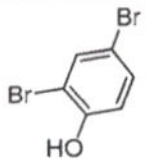	
Bmp2 [127]	*Pseudoalteromonas luteoviolacea*		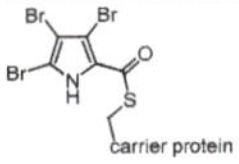	NRPS-dependent
(PltD) [13]	*Pseudomonas fluorescens* Pf-5		n.a.	Not clear if FHal
CmlS [128]	*Streptomyces venezuelae*		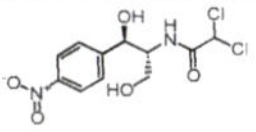	Flavin covalently bound to aspartate via CH_3-Group
CndH [129]	*Chondromyces crocatus*		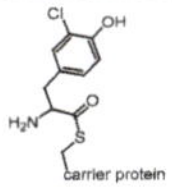	NRPS-dependent
RadH [107,130]	*Chaetomium chiversii*	*E. coli* Rosetta 2(DE3)	monocillin II	
Rdc2 [107,131]	*Pochonia chlamydosporia*	*S. cerevisiae* strain BJ5464-Npg *E. coli* BL21(DE3)	monocillin II	
TiaM [132]	*Dactylosporangium aurantiacum* NRRL 18085	*E. coli* BL21(DE3)	tiacumicin B intermediate	NRPS-dependent

[1] if not stated otherwise, the expression took place in the origin host.

5.2. *α-Ketoglutarate-Dependent Halogenases*

Table 4 shows different natural products that are formed by the iron(II)-α-ketoglutarate-dependent (Fe/αKG)-halogenases. Despite the huge variety in the product structures they share one common feature, which is the halogen at a sp^3-carbon centre. Hence, the Fe/αKG-halogenase is not limited to nucleophilic substrates like the previous described enzymes. They belong to the Fe/αKG-dependent oxygenase superfamily. The superfamily is known for different transformations such as hydroxylation [138], halogenation [139], desaturation [140], or can be used for the production of ethylene [141]. They all share a structurally conserved metal-binding motif, which in the case of the halogenase developed an active centre that is eventually able to bind a haloge n [139]. The proposed catalytic mechanism of Fe(II)/α-KG-dependent-halogenase is illustrated in Figure 16.

Table 4. Examples of heterologously expressed Fe(II)/αKG-dependent halogenases.

Enzyme	Origin/Expression Host and Yield	Features	Product/Biosynthesis
SyrB2 [139]	*Pseudomonas syringae pv. syringae* B301D/ *E. coli* strain B834(DE3) [139] n.a.	total turnover: 7 ± 2 [144]	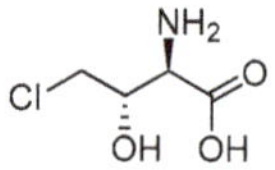 4-chloro-L-threonine/ syringomycin E
CytC3 [145]	*Streptomyces* sp./ *E. coli* BL21(DE3) [146] n.a.	n.a	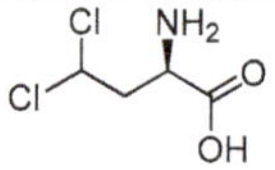 4,4-dichloro-L-valine/ dichloroaminobutyrate
WelO5 [147]	*Hapalosiphon welwitschii*/ *E. coli* C43(DE3) [148,149] or BL21(DE3) [150] 20 mg L^{-1} [151]	total turnover: 75 [148] k_{cat}: 1.8 ± 0.2 min^{-1} [149]	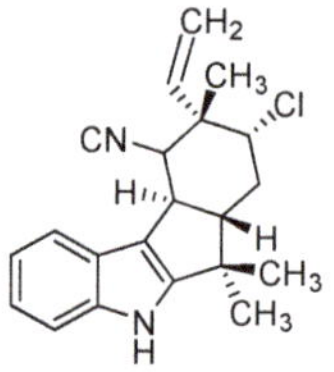 12-*epi*-fischerindole G/ fischerindole & hapalindole alkaloids
AmbO5	*Fischerella ambigua*/ *E. coli* C43(DE3) [149] or BL21(DE3) [150] n.a.	k_{cat}: 1.7 ± 0.1 min^{-1} [149]	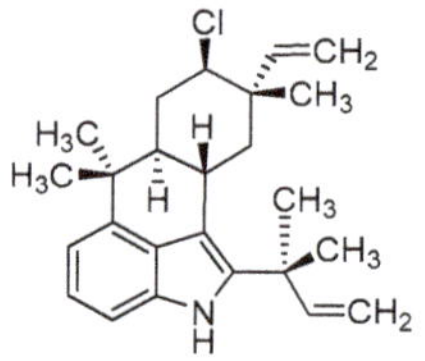 ambiguine A/ ambiguine, fischerindole and hapalindole alkaloids
WelO5* variant isoform of WelO5 (CB2) [152]	*Hapalosiphon welwitschii* IC-52-3/ *E. coli* BL21(DE3) n.a.	K_M: 0.67 mM, k_{cat}: 3.0 min^{-1}	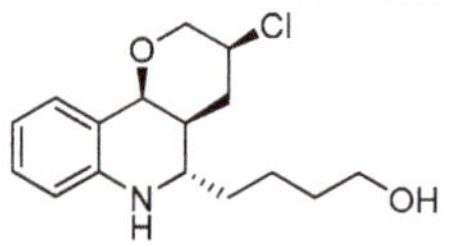 martinelline-derived fragment

Based on the proposed radical C-H functionalization two classes of enzymes have so far been identified. The first such reported Fe/αKG-dependent halogenase is the tailoring domain SyrB2 of the multimodular nonribosomal peptidsynthetase (NRPS) from *Pseudomonas syringae* pv. *syringae* B301D [133,144]. These NRPS-associated halogenases produce a diversity of secondary metabolites such as the chlorinated biosurfactant syringomycin E (**31**), which is characterized by a selective monochlorinated threonine in its structure [128,139]. Another example is the highly selective di- and trichlorination of solely one of the diastereotopic methyl groups of leucine by a combination of BarB1 and BarB2, which serves as a precursor for the natural compound barbamid (**32**) in the marine cyanobacteria *Lyngbya majuscula* [142,153].

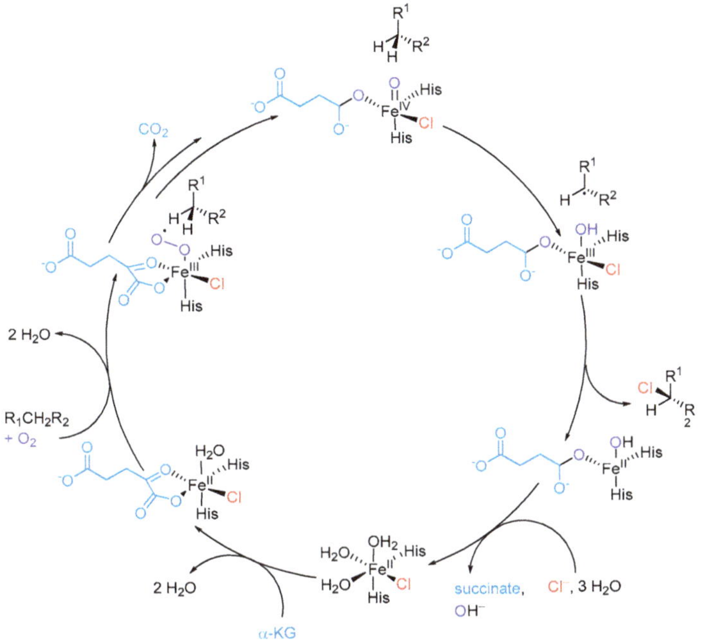

Figure 16. Proposed mechanism for halogenation reaction by Fe(II)/αKG-dependent halogenase via a radical C-H functionalization [142]. The highly reactive Fe(IV)-oxo (haloferryl) intermediate is produced by decarboxylation of αKG to succinate through an oxygen attack. Subsequently abstraction of a hydrogen-atom from the substrate leads to an energetically favourable rearrangement towards Fe(III). Rebound reaction with chloride was shown to depend on the distance and orientation of the substrate [143]. The catalytically cycle is re-established by the hexa-coordinated Fe(II) with water molecules, chloride and histidine.

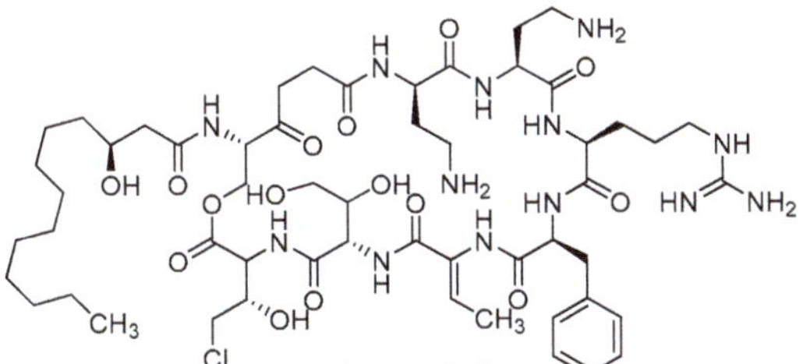

syringomycin E
31

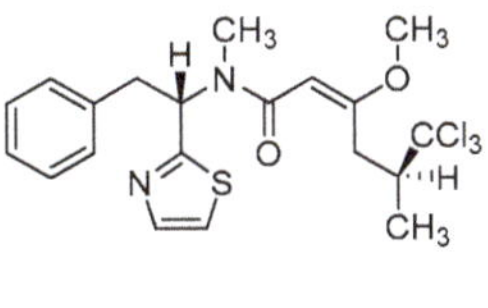

barbamid
32

Recently Moosmann et al. identified different αKG-halogenase homologues and their natural products that are produced via a NRPS (non-ribosomal peptide synthetase) pathway. The halogenase were identified by screening the genomic sequence of the cyanobacterium *Fischerella* sp. PCC 9339 based on feature comparison. Using this approach, the authors were able to distinguish between a Fe/αKG oxygenase and a corresponding halogenase [143,154]. However, large NRPSs characteristically bind their substrates through an aminoacetylated peptidyl-carrier protein and have a narrow substrate scope [144,155]. Furthermore, they generally showed a low total turnover number, which may result from the well-known autoxidation of Fe(II) to Fe(III) and hence an auto-inactivation of the enzyme [145,156]. In case of SyrB2, total turnovers of 7 ± 2 were observed [133,144] This limits the possibility to modify the enzymes in order to use them as suitable biocatalysts for different unnatural substrates. With the discovery of a new Fe(II)/αKG-dependent halogenase (WelO5) by Hillwig and Liu, it was possible to expand the class towards unbound substrates. WelO5 is capable of late stage halogenation in a regio- and stereoselective manner of different derived isoprenoid-indole alkaloids in the cyanobacterium *Hapalosiphon welwitschii* (see Table 4) [137,148]. WelO5 showed also a higher robustness and catalysis of approximately 75 turnovers in total [137,148]. Strategies such as adding the cosubstrates consecutively or adding antioxidants like catalases or DTT could increase the turnover number. The narrow substrate scope of WelO5 was tailored in order to have an increased substrate scope like the homolog AmbO5 [138,149]. Most modifications were at the external helix, which is responsible for closing the entry of the active site upon binding of the substrate. It can be assumed that the helix is partially involved in the substrate recognition and specificity [138,149]. A recent publication from Hayashi et al. showed a WelO5 variant with a reshaped active site that led to improved kinetics and an expanded substrate scope, which applies beyond the native indole alkaloid-type substrates [141,152]. This provides the possibility for targeted enzyme-engineering and a basis for further improvements in substrate scope. One possibility is the establishment of nitration and azidation as already shown for SyrB2 [157]. In this regard, it has been shown that WelO5 is able to incorporate the unnatural halide Br^- [158]. One drawback of engineering Fe/αKG-dependent halogenases is the hydroxylation as a competitive side reaction [155]. Mitchell et al. used this approach backwards and modified a monooxygenase SadA towards a halogenase [159]. This serves as a proof of concept that with increasing understanding of the reaction mechanism and the involved amino acids the superfamily of monooxygenase can be used as a versatile toolbox in biotechnology. In the future, this may lead to the use of different variants of the very same enzyme for different transformations. Table 4 shows an overview of different characterized Fe/αKG halogenases and their main published features. Excluded are, for example, halogenase modules of NRPS, where the halogenation is necessary for the subsequently formation of cyclopropane such as in case of CurA [160] or CmaB (see Figure 17) [161].

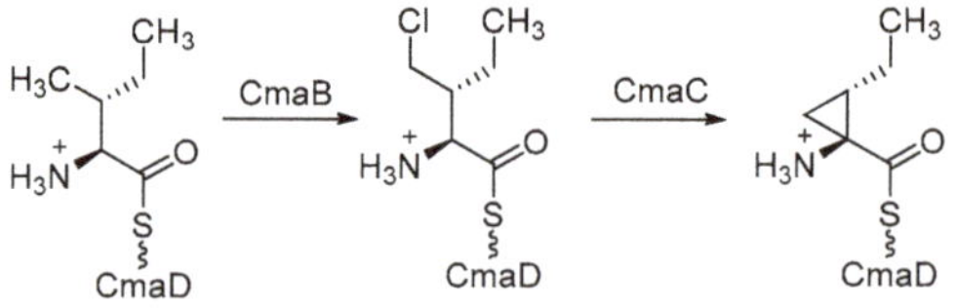

Figure 17. Schematically example for formation of cyclopropane initiated by CmaB through halogenation.

5.3. Fluorinase

In contrast to the other described enzymes, the diversity of natural products in case of the fluorinase stem from only one characterized enzyme class to date. The involved enzymes are *S*-adenosylmethionine (SAM) dependent. The first characterized representative was FlA (5′-fluoro-5′-deoxyadenosine synthase) from [162]. The overall family of this enzymes is also able to chlorinate or hydroxylate SAM, as described in detail elsewhere [163]. Within the catalytic cycle fluoride acts as a nucleophile in a S_N2-reaction, where it attacks the 5′-carbon of SAM-ribose [164]. In order to act as a nucleophile, fluoride requires to lose its solvation shell. This is achieved in a two-step desolvation with a combination of electrostatic stabilization and hydrogen bonding. In the first step, fluoride is binding to the active site and exchanges water molecules of its shell in order to form hydrogen bonds with the enzyme. Upon binding of SAM the desolvation of fluoride is complete. The electropositive 5′-carbon attached to the sulfur group in SAM coordinates with the fluoride [165,166]. This electrostatic stabilization facilitates the nucleophilic attack of the fluoride and C–F-bond formation of the reactive 5′-fluoro-5′deoxyadenosine (**33**, 5′-FDA) intermediate [166]. Subsequently, 5′-FDA (**33**) is further metabolized in order to generate a variety of compounds as shown in Figure 18A [162]. However, this also represents a major obstacle for the application of these enzymes to unnatural small organic molecules, since the product formation follows a cascade of enzymatic steps. Eustáquio et al. tried to use this enzyme for the production of fluorosalinosporamide, an unnatural analog of salinosporamide, which is fluorinated rather than chlorinated, however, the yield was moderate [167]. Nevertheless, different approaches have been implemented to increase the substrate scope and use the enzyme as rather flexible tool for medical applications. Besides the ability to fluoride compounds, fluorinase is also able to exchange a chloride at the 5′-carbon of the SAM ribose ring by a fluoride and form 5′-FDA (**33**, Figure 18B) [168]. This overall trans-halogenation reaction was used for late-stage fluorination for the production of radiolabeled imaging reagents. Recently, different pre-targeting strategies have been developed for treatment and imaging of different diseases. Those include e.g., radiolabeling of the human A2A adenosine receptor [169], a prostate cancer-related membrane protein [170] or the combined application of biotin and tetrazine-conjugate with antibodies [171]. In all cases, it was shown that the fluorinase (FlA) accepts substrates with different moiety at C-2 of the adenine ring. So far two crystal structures of fluorinase homologs from *S. cattleya* and *Streptomyces* sp. MA37 are known and they both have a high structurally conformity [172]. In general all five known fluorinases have a high similarity of over 80% and show similar kinetic profiles [173]. Through a directed evolution approach of FlA1, different crucial amino acids for substrate binding, halide binding and hence activity were identified [174]. Additionally, the variants were tested with different unnatural substrates [175]. It was shown that the tolerance for the *wild type* (*wt*) enzyme is limited to C-2 modified substrates. However, generated variants of FlA1 also demonstrated an activity for an unnatural substrate, which were modified at C-6 positions of the adenine ring with a chlorine group [175]. These findings show that despite a narrow substrate scope of the fluorinase, it was possible to successfully apply different unnatural substrates and lay a base for directed evolution as means to use small organic compounds as substrates. However, the dependence of electrophilic substrate structures remains a drawback for the nucleophilic attack of fluoride. The crystal structure with an unnatural substrate (containing difluoromethyl groups) confirmed the necessity of geometry for activating the fluorine atom for substitution [176]. This outlines the challenge to use fluorinases as a versatile tool to generate fluorinated pharmaceutical compounds. Nevertheless, by means of designing appropriate leaving groups in combination with enzyme engineering, fluorinases could be used as tool for future generation of fluorinated pharmaceutical compounds. Data of known fluorinases are displayed in Table 5.

Table 5. Examples for heterologously expressed fluorinase. Kinetic data representing the conversion of 5′-ClDA into *S*-adenosylmethionine (SAM).

Enzyme	Origin/Expression Host and Yield	Kinetic Parameters	Special Substrate Scope
FlA [165]	*Streptomyces cattleya*/ *E. coli* BL21(DE3) 50 mg L^{-1} [173]	K_M: 29.4 ± 5.80 µM k_{cat}: 0.084 ± 0.005 min^{-1} [173]	[169–171]
FlA1	*Streptomyces* sp. MA37/ *E. coli* BL21(DE3) n.a.	K_M: 8.36 ± 0.82 µM k_{cat}: 0.13 min^{-1} [174]	
FlA4	*Streptomyces xinghaiensis*/ *E. coli* BL21(DE3) n.a.	K_M: 29.87 µM k_{cat}: 0.69 ± 0.01 min^{-1} [177]	[177]

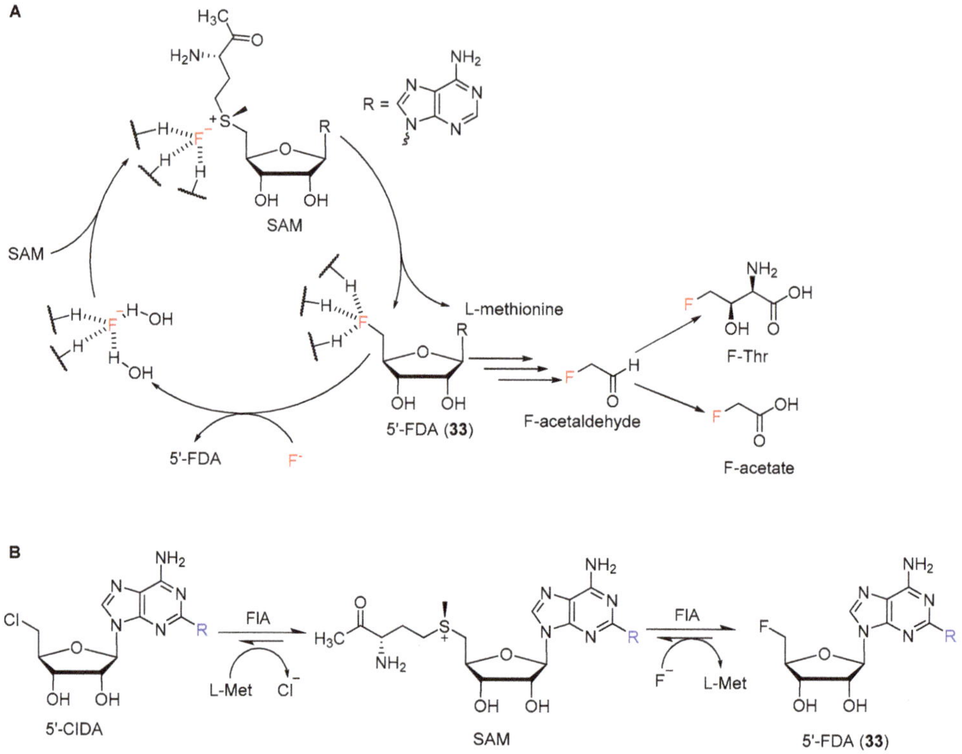

Figure 18. (**A**) Schematically sequential mechanism for the F–C bond formation catalysed by the fluorinase from *S. cattleya* and some products of subsequently cascade reaction. Dashed lines represents hydrogen bonding contacts with amino acids in the active pocket or water. (**B**) Reaction scheme of fluorinase-mediated trans-halogenation. Rest R marks position of usual derivation.

6. Conclusion on Halogens in Active Agent (Syntheses)

As we have seen in the previous paragraphs halogens are very important to many active agents as a functional moiety *per se* due to their physico-chemical properties such as bulkiness, latent polarization and as important binding partners because of halogen bonds. Organic halogen compounds are, furthermore, instrumental for synthetic purposes in terms of being good leaving groups and facilitating cross-metatheses by halogen-metal-exchanges. Nevertheless, these indispensable advantages have to be bought at a high price; namely the energy-intensive production of very toxic and hazardous chlorine gas. The reduction in energy consumption must mainly be managed by technical improvements of the chloralkali-process and enzymes are likely not able to make a significant impact. A major reason is that the majority of chlorinated compounds are necessary for different types of plastic materials (e.g., PVC) and solvents [51]. Enzymes are limitedly applicable in those areas of bulk chemicals, but there is a potential for fine chemicals. Even though halogenating enzymes will not replace conventional chlorine production, it is worth taking a look at this group of enzymes or rather at these groups of enzymes, because nature has invented these amazing enzymes at least six times. The expectations of these biocatalysts are that the conversions become environmentally more benign, the processes skip hazardous compounds such as chlorine gas and that conversions get more selective. However, the research in the field of halogenating enzymes is still at the beginning. Consistent enzymologic data such as kinetic data, measurements on stability or even well studied mutant libraries are rarely available. Many halogenating enzymes from eukaryotic sources suffer from expression challenges. Nevertheless, these enzymes open up a wide horizon of possibilities. Enormous genome data are revealing more and more halogenating enzymes and even new classes of halogenating enzymes cannot be excluded at present. Thus, there is a need for detailed and systematic research to employ halogenating enzymes for active agent synthesis, to alter their substrate scopes and enhance their process stability.

Author Contributions: T.C. conceived this review and relegated the manuscript to a complete work. A.V.F., J.G., and N.H. carried out the initial research and compiled the various chapters. All authors contributed equally to this review.

Funding: This research was funded by Deutsche Forschungsgemeinschaft grant number 369034981, BioSC(CombiCom) Ministry of Culture and Science North-Rhine-Wesfalia grant number 313/323-400-00213, and European Regional Development Fund: CLIB-Kompetenzzentrum Biotechnologie.

Acknowledgments: For their fruitful discussions we want to thank *Jörg Pietruszka*, *Nora Bitzenhofer*, *Esther Breuninger*, and *Denise Detlof* as well as the entire institute staff for their support. We want to thank our sponsors for the financial support. AF was funded by the *European Regional Development Fund (ERDF)* within the '*CLIB-Kompetenzzentrum Biotechnologie*'. The scientific activities (JG) of the *Bioeconomy Science Center* were financially supported by the *Ministry of Culture and Science* within the framework of the *NRW Strategieprojekt BioSC* (No. 313/323-400-00213). NH is funded by the *Deutsche Forschungsgemeinschaft* (369034981). Furthermore, we want to thank the *Forschungszentrum Jülich GmbH* and the *Heinrich Heine University Düsseldorf* for their generous support.

Conflicts of Interest: The authors declare no conflict of interest.

Abbreviations

αKG	α-ketoglutarate
BPO	bromoperoxidase
CLEA	cross-linked enzyme aggregate
CPO	chloroperoxidase
FDA	5′-fluoro-5′deoxyadenosine
Fhal, FDH	flavin-dependent halogenase
FlA	fluorinase A
g	gasous
Hal	halogen
HOX	hypohalous acid
HPO	haloperoxidase
MCD	monochlorodimedone
Mn	elemental metal
MOH	metal cation hydroxid salts
NBS	*N*-bromo-succinimde
NCS	*N*-chloro-succinimide
NRPS	non-ribosomal protein synthesis
NXS	*N*-halogen-succinimide
PKS	polyketide
PVC	polyvinyl chlroide
s	solid
SAM	*S*-adenosyl methionine
X	halide ion
X_2	molecular halogen
Y	non-halogen heteroatom

References

1. Gribble, G.W. Natural Organohalogens: A New Frontier for Medicinal Agents? *J. Chem. Educ.* **2004**, *81*, 1441. [CrossRef]
2. Chast, F. A history of drug discovery: From first steps of chemistry to achievements in molecular pharmacology. In *The Practice of Medicinal Chemistry*; Elsevier: Amsterdam, The Netherlands, 2008; pp. 1–62.

3. Hernandes, M.Z.; Cavalcanti, S.M.T.; Moreira, D.R.M.; de Azevedo Junior, W.F.; Leite, A.C.L. Halogen Atoms in the Modern Medicinal Chemistry: Hints for the Drug Design. *Curr. Drug Targets* **2010**, *11*, 303–314. [CrossRef] [PubMed]
4. Jiang, S.; Zhang, L.; Cui, D.; Yao, Z.; Gao, B.; Lin, J.; Wei, D. The Important Role of Halogen Bond in Substrate Selectivity of Enzymatic Catalysis. *Sci. Rep.* **2016**, *6*, 34750. [CrossRef] [PubMed]
5. Agarwal, V.; Miles, Z.D.; Winter, J.M.; Eustáquio, A.S.; EL Gamal, A.A.; Moore, B.S. Enzymatic Halogenation and Dehalogenation Reactions: Pervasive and Mechanistically Diverse. *Chem. Rev.* **2017**, *117*, 5619–5674. [CrossRef]
6. Blasiak, L.C.; Drennan, C.L. Structural perspective on enzymatic halogenation. *Acc. Chem. Res.* **2008**, *42*, 147–155. [CrossRef]
7. Gribble, G.W. The diversity of naturally produced organohalogens. *Chemosphere* **2003**, *52*, 289–297. [CrossRef]
8. Latham, J.; Brandenburger, E.; Shepherd, S.A.; Menon, B.R.K.; Micklefield, J. Development of Halogenase Enzymes for Use in Synthesis. *Chem. Rev.* **2018**, *118*, 232–269. [CrossRef]
9. Schnepel, C.; Sewald, N. Enzymatic Halogenation: A Timely Strategy for Regioselective C−H Activation. *Chem. Eur. J.* **2017**, *23*, 12064–12086. [CrossRef]
10. Senn, H.M. Insights into enzymatic halogenation from computational studies. *Front. Chem.* **2014**, *2*, 98. [CrossRef]
11. Smith, D.R.; Grüschow, S.; Goss, R.J. Scope and potential of halogenases in biosynthetic applications. *Curr. Opin. Chem. Biol.* **2013**, *17*, 276–283. [CrossRef] [PubMed]
12. Vaillancourt, F.H.; Yeh, E.; Vosburg, D.A.; Garneau-Tsodikova, S.; Walsh, C.T. Nature's inventory of halogenation catalysts: Oxidative strategies predominate. *Chem. Rev.* **2006**, *106*, 3364–3378. [CrossRef] [PubMed]
13. Weichold, V.; Milbredt, D.; van Pée, K.-H. Specific Enzymatic Halogenation—From the Discovery of Halogenated Enzymes to Their Applications In Vitro and In Vivo. *Angew. Chem. Int. Ed.* **2016**, *55*, 6374–6389. [CrossRef] [PubMed]
14. Zeng, J.; Zhan, J. Chlorinated Natural Products and Related Halogenases. *Isr. J. Chem.* **2019**, *59*, 1–17. [CrossRef]
15. Lu, Y.; Wang, Y.; Zhu, W. Nonbonding interactions of organic halogens in biological systems: Implications for drug discovery and biomolecular design. *Phys. Chem. Chem. Phys.* **2010**, *12*, 4543–4551. [CrossRef] [PubMed]
16. Xu, Z.; Yang, Z.; Liu, Y.; Lu, Y.; Chen, K.; Zhu, W. Halogen bond: Its role beyond drug–target binding affinity for drug discovery and development. *J. Chem. Inf. Model.* **2014**, *54*, 69–78. [CrossRef] [PubMed]
17. Sirimulla, S.; Bailey, J.B.; Vegesna, R.; Narayan, M. Halogen interactions in protein–ligand complexes: Implications of halogen bonding for rational drug design. *J. Chem. Inf. Model.* **2013**, *53*, 2781–2791. [CrossRef] [PubMed]
18. Lu, Y.; Liu, Y.; Xu, Z.; Li, H.; Liu, H.; Zhu, W. Halogen bonding for rational drug design and new drug discovery. *Expert Opin. Drug Discov.* **2012**, *7*, 375–383. [CrossRef] [PubMed]
19. Xu, Z.; Liu, Z.; Chen, T.; Chen, T.; Wang, Z.; Tian, G.; Shi, J.; Wang, X.; Lu, Y.; Yan, X.; et al. Utilization of halogen bond in lead optimization: A case study of rational design of potent phosphodiesterase type 5 (PDE5) inhibitors. *J. Med. Chem.* **2011**, *54*, 5607–5611. [CrossRef]
20. Harris, C.M.; Kannan, R.; Kopecka, H.; Harris, T.M. The role of the chlorine substituents in the antibiotic vancomycin: Preparation and characterization of mono-and didechlorovancomycin. *J. Am. Chem. Soc.* **1985**, *107*, 6652–6658. [CrossRef]
21. Schlosser, M.; Michel, D. About the "physiological size" of fluorine substituents: Comparison of sensorially active compounds with fluorine and methyl substituted analogues. *Tetrahedron* **1996**, *52*, 99–108. [CrossRef]
22. Lima, L.M.; Barreiro, E.J. Bioisosterism: A useful strategy for molecular modification and drug design. *Curr. Med. Chem.* **2005**, *12*, 23–49. [CrossRef]
23. Greenwood, N.N.; Earnshaw, A. *Chemistry of the Elements*; Elsevier: Amsterdam, The Netherlands, 2012.
24. Thiehoff, C.; Rey, Y.P.; Gilmour, R. The Fluorine *Gauche* Effect: A Brief History. *Isr. J. Chem.* **2017**, *57*, 92–100. [CrossRef]
25. Berkowitz, D.B.; Karukurichi, K.R.; de la Salud-Bea, R.; Nelson, D.L.; McCune, C.D. Use of fluorinated functionality in enzyme inhibitor development: Mechanistic and analytical advantages. *J. Fluor. Chem.* **2008**, *129*, 731–742. [CrossRef]

26. Ramalingam, S.S.; Blackhall, F.; Krzakowski, M.; Barrios, C.H.; Park, K.; Bover, I.; Heo, D.S.; Rosell, R.; Talbot, D.C.; Frank, R. Randomized phase II study of dacomitinib (PF-00299804), an irreversible pan–human epidermal growth factor receptor inhibitor, versus erlotinib in patients with advanced non–small-cell lung cancer. *J. Clin. Oncol.* **2012**, *30*, 3337. [CrossRef]
27. Bush, J.A.; Long, B.H.; Catino, J.J.; Bradner, W.T.; Tomita, K. Production and biological activity of rebeccamycin, a novel antitumor agent. *J. Antibiot.* **1987**, *40*, 668–678. [CrossRef]
28. Lake, R.J.; Brennan, M.M.; Blunt, J.W.; Munro, M.H.; Pannell, L.K. Eudistomin K sulfoxide-an antiviral sulfoxide from the New Zealand ascidian *Ritterella sigillinoides*. *Tetrahedron Lett.* **1988**, *29*, 2255–2256. [CrossRef]
29. Braghirolli, A.M.S.; Waissmann, W.; da Silva, J.B.; dos Santos, G.R. Production of iodine-124 and its applications in nuclear medicine. *Appl. Radiat. Isot.* **2014**, *90*, 138–148. [CrossRef]
30. Crawford, C.C.; Evans, T.W. Production of Hypohalous Acid Solutions. U.S. Patent 2347151, 18 April 1944.
31. Duncan, B.L.; Ness, R.C. Process for the Production of Highly Pure Concentrated Slurries of Sodium Hypochlorite. U.S. Patent 5194238A, 16 March 1993.
32. Howarth, J.N.; Dadgar, A.; Sergent, R.H. Recovery of Bromine and Preparation of Hypobromous acid from Bromide Solution. U.S. Patent 5385650A, 31 January 1995.
33. Kesner, M. *Bromine and Bromine Compounds from the Dead Sea, Israel Products in the Service of People*; The Weizmann Institute of Science, Israel: Rehovot, Israel, 1999.
34. Groweiss, A. Use of Sodium Bromate for Aromatic Bromination: Research and Development. *Org. Process Res. Dev.* **2000**, *4*, 30–33. [CrossRef]
35. Nishi, Z.; Kudoh, K.; Okamoto, N. Method of Forming Hypobromous Acid in Aqueous System. U.S. PATENT 7785559B2, 31 August 2010.
36. Ziegler, K.; Späth, A.; Schaaf, E.; Schumann, W.; Winkelmann, E. Die halogenierung ungesättigter Substanzen in der Allylstellung. *Justus Liebigs Ann. Chem.* **1942**, *551*, 80–119. [CrossRef]
37. Torborg, C.; Beller, M. Recent applications of palladium-catalyzed coupling reactions in the pharmaceutical, agrochemical, and fine chemical industries. *Adv. Synth. Catal.* **2009**, *351*, 3027–3043. [CrossRef]
38. Beletskaya, I.P.; Cheprakov, A.V. Copper in cross-coupling reactions: The post-Ullmann chemistry. *Coord. Chem. Rev.* **2004**, *248*, 2337–2364. [CrossRef]
39. Xi, Z.; Liu, B.; Chen, W. Room-temperature Kumada cross-coupling of unactivated aryl chlorides catalyzed by *N*-heterocylic carbene-based nickel (II) complexes. *J. Org. Chem.* **2008**, *73*, 3954–3957. [CrossRef]
40. Gomes, F.; Echeverria, P.G.; Fürstner, A. Iron-or Palladium-Catalyzed Reaction Cascades Merging Cycloisomerization and Cross-Coupling Chemistry. *Chem. Eur. J.* **2018**, *24*, 16814–16822. [CrossRef]
41. Piontek, A.; Bisz, E.; Szostak, M. Iron-Catalyzed Cross-Couplings in the Synthesis of Pharmaceuticals: In Pursuit of Sustainability. *Angew. Chem. Int. Ed.* **2018**, *57*, 11116–11128. [CrossRef]
42. Cahiez, G.R.; Lefèvre, G.; Moyeux, A.; Guerret, O.; Gayon, E.; Guillonneau, L.; Lefèvre, N.; Gu, Q.; Zhou, E. Gram-Scale, Cheap, and Eco-Friendly Iron-Catalyzed Cross-Coupling between Alkyl Grignard Reagents and Alkenyl or Aryl Halides. *Org. Lett.* **2019**, *21*, 2679–2683. [CrossRef]
43. Cooke, J.W.; Bright, R.; Coleman, M.J.; Jenkins, K.P. Process research and development of a dihydropyrimidine dehydrogenase inactivator: Large-scale preparation of eniluracil using a Sonogashira coupling. *Org. Process Res. Dev.* **2001**, *5*, 383–386. [CrossRef]
44. Wallace, M.D.; McGuire, M.A.; Yu, M.S.; Goldfinger, L.; Liu, L.; Dai, W.; Shilcrat, S. Multi-kiloscale enantioselective synthesis of a vitronectin receptor antagonist. *Org. Process Res. Dev.* **2004**, *8*, 738–743. [CrossRef]
45. Benkeser, R.A.; Siklosi, M.P.; Mozdzen, E.C. Reversible Grignard and organolithium reactions. *J. Am. Chem. Soc.* **1978**, *100*, 2134–2139. [CrossRef]
46. Milstein, D.; Stille, J.K. A general, selective, and facile method for ketone synthesis from acid chlorides and organotin compounds catalyzed by palladium. *J. Am. Chem. Soc.* **1978**, *100*, 3636–3638. [CrossRef]
47. King, A.O.; Okukado, N.; Negishi, E.-I. Highly general stereo-, regio-, and chemo-selective synthesis of terminal and internal conjugated enynes by the Pd-catalysed reaction of alkynylzinc reagents with alkenyl halides. *Chem. Commun.* **1977**, 683–684. [CrossRef]
48. Corriu, R.; Masse, J. Activation of Grignard reagents by transition-metal complexes. A new and simple synthesis of trans-stilbenes and polyphenyls. *Chem. Commun.* **1972**, *144*. [CrossRef]

49. Global Chlorine Market Set for Rapid Growth to Reach around USD 38.4 Billion in 2021. Available online: https://www.zionmarketresearch.com/news/global-chlorine-market (accessed on 23 August 2019).
50. Ausfelder, F. *Flexibilitätsoptionen in der Grundstoffindustrie: Methodik, Potenziale, Hemmnisse: Bericht des AP V. 6 "Flexibilitätsoptionen und Perspektiven in der Grundstoffindustrie "im Kopernikus-Projekt "SynErgie-Synchronisierte und Energieadaptive Produktionstechnik zur Flexiblen Ausrichtung von Industrieprozessen Auf Eine Fluktuierende Energieversorgung"*; DECHEMA Gesellschaft für Chemische Technik und Biotechnologie eV: Frankfurt, Germany, 2018.
51. Chlor-Alkali Industry Review 2017/2018. Available online: https://www.eurochlor.org/wp-content/uploads/2019/05/euro_chlor_industry_review_FINAL.pdf (accessed on 30 October 2019).
52. The Electrolysis Process and the Real Costs of Production. Available online: https://www.eurochlor.org/wp-content/uploads/2019/04/12-electrolysis_production_costs.pdf (accessed on 23 August 2019).
53. Morris, D.R.; Hager, L.P.; Chloroperoxidase, I. Isolation and properties of the crystalline glycoprotein. *J. Biol. Chem.* **1966**, *241*, 1763–1768. [PubMed]
54. Kuhnel, K.; Blankenfeldt, W.; Terner, J.; Schlichting, I. Crystal structures of chloroperoxidase with its bound substrates and complexed with formate, acetate, and nitrate. *J. Biol. Chem.* **2006**, *281*, 23990–23998. [CrossRef] [PubMed]
55. Butler, A.; Sandy, M. Mechanistic considerations of halogenating enzymes. *Nature* **2009**, *460*, 848. [CrossRef] [PubMed]
56. Buchhaupt, M.; Huttmann, S.; Sachs, C.C.; Bormann, S.; Hannappel, A.; Schrader, J. *Caldariomyces fumago* DSM1256 Contains Two Chloroperoxidase Genes, Both Encoding Secreted and Active Enzymes. *J. Mol. Microbiol. Biotechnol.* **2015**, *25*, 237–243. [CrossRef] [PubMed]
57. Reddy, C.M.; Xu, L.; Drenzek, N.J.; Sturchio, N.C.; Heraty, L.J.; Kimblin, C.; Butler, A. A chlorine isotope effect for enzyme-catalyzed chlorination. *J. Am. Chem. Soc.* **2002**, *124*, 14526–14527. [CrossRef]
58. Pickard, M.A. A defined growth medium for the production of chloroperoxidase by *Caldariomyces fumago*. *Can. J. Microbiol.* **1981**, *27*, 1298–1305. [CrossRef]
59. Getrey, L.; Krieg, T.; Hollmann, F.; Schrader, J.; Holtmann, D. Enzymatic halogenation of the phenolic monoterpenes thymol and carvacrol with chloroperoxidase. *Green Chem.* **2014**, *16*, 1104–1108. [CrossRef]
60. Naapuri, J.; Rolfes, J.D.; Keil, J.; Manzuna Sapu, C.; Deska, J. Enzymatic halocyclization of allenic alcohols and carboxylates: A biocatalytic entry to functionalized *O*-heterocycles. *Green Chem.* **2017**, *19*, 447–452. [CrossRef]
61. Blanke, S.R.; Yi, S.; Hager, L.P. Development of Semi-Continuous and Continuous-Flow Bioreactors for the High-Level Production of Chloroperoxidase. *Biotechnol. Lett.* **1989**, *11*, 769–774. [CrossRef]
62. Yamada, H.; Itoh, N.; Izumi, Y. Chloroperoxidase-catalyzed halogenation of *trans*-cinnamic acid and its derivatives. *J. Biol. Chem.* **1985**, *260*, 11962–11969. [PubMed]
63. Allain, E.J.; Hager, L.P.; Deng, L.; Jacobsen, E.N. Highly Enantioselective Epoxidation of Disubstituted Alkenes with Hydrogen-Peroxide Catalyzed by Chloroperoxidase. *J. Am. Chem. Soc.* **1993**, *115*, 4415–4416. [CrossRef]
64. Manthey, J.A.; Hager, L.P. Purification and properties of bromoperoxidase from *Penicillus capitatus*. *J. Biol. Chem.* **1981**, *256*, 11232–11238. [PubMed]
65. van Pée, K.H.; Lingens, F. Purification of bromoperoxidase from *Pseudomonas aureofaciens*. *J. Bacteriol.* **1985**, *161*, 1171–1175.
66. Baden, D.G.; Corbett, M.D. Bromoperoxidases from Penicillus capitatus, Penicillus lamourouxii and Rhipocephalus phoenix. *Biochem. J.* **1980**, *187*, 205–211. [CrossRef]
67. Kaup, B.-A.; Ehrich, K.; Pescheck, M.; Schrader, J. Microparticle-enhanced cultivation of filamentous microorganisms: Increased chloroperoxidase formation by *Caldariomyces fumago* as an example. *Biotechnol. Bioeng.* **2008**, *99*, 491–498. [CrossRef]
68. Conesa, A.; van de Velde, F.; van Rantwijk, F.; Sheldon, R.A.; van den Hondel, C.A.M.J.J.; Punt, P.J. Expression of the *Caldariomyces fumago* chloroperoxidase in *Aspergillus niger* and characterization of the recombinant enzyme. *J. Biol. Chem.* **2001**, *276*, 17635–17640. [CrossRef]
69. Coughlin, P.; Roberts, S.; Rush, C.; Willetts, A. Biotransformation of Alkenes by Haloperoxidases - Regiospecific Bromohydrin Formation from Cinnamyl Substrates. *Biotechnol. Lett.* **1993**, *15*, 907–912. [CrossRef]

70. van Schijndel, J.W.; Vollenbroek, E.G.; Wever, R. The chloroperoxidase from the fungus *Curvularia inaequalis*; a novel vanadium enzyme. *Biochim. Biophys. Acta* **1993**, *1161*, 249–256. [CrossRef]
71. Van Schijndel, J.W.P.M.; Barnett, P.; Roelse, J.; Vollenbroek, E.G.M.; Wever, R. The stability and steady-state kinetics of vanadium chloroperoxidase from the fungus *Curvularia inaequalis*. *Eur. J. Biochem.* **1994**, *225*, 151–157. [CrossRef]
72. Rush, C.; Willetts, A.; Davies, G.; Dauter, Z.; Watson, H.; Littlechild, J. Purification, crystallisation and preliminary X-ray analysis of the vanadium-dependent haloperoxidase from *Corallina officinalis*. *FEBS Lett.* **1995**, *359*, 244–246. [CrossRef]
73. Wever, R. *Vanadium: Biochemical and Molecular Biological Approaches*; Springer: New York, NY, USA, 2012.
74. Leblanc, C.; Vilter, H.; Fournier, J.B.; Delage, L.; Potin, P.; Rebuffet, E.; Michel, G.; Solari, P.L.; Feiters, M.C.; Czjzek, M. Vanadium haloperoxidases: From the discovery 30 years ago to X-ray crystallographic and V K-edge absorption spectroscopic studies. *Coord. Chem. Rev.* **2015**, *301–302*, 134–146. [CrossRef]
75. Hemrika, W.; Renirie, R.; Macedo-Ribeiro, S.; Messerschmidt, A.; Wever, R. Heterologous Expression of the Vanadium-containing Chloroperoxidase from *Curvularia inaequalis* in *Saccharomyces cerevisiae* and Site-directed Mutagenesis of the Active Site Residues His496, Lys353, Arg360, and Arg490. *J. Biol. Chem.* **1999**, *274*, 23820–23827. [CrossRef] [PubMed]
76. Fernandez-Fueyo, E.; van Wingerden, M.; Renirie, R.; Wever, R.; Ni, Y.; Holtmann, D.; Hollmann, F. Chemoenzymatic Halogenation of Phenols by using the Haloperoxidase from *Curvularia inaequalis*. *ChemCatChem* **2015**, *7*, 4035–4038. [CrossRef]
77. Dong, J.J.; Fernandez-Fueyo, E.; Li, J.; Guo, Z.; Renirie, R.; Wever, R.; Hollmann, F. Halofunctionalization of alkenes by vanadium chloroperoxidase from *Curvularia inaequalis*. *Chem. Commun.* **2017**, *53*, 6207–6210. [CrossRef] [PubMed]
78. Fernández-Fueyo, E.; Younes, S.H.; Rootselaar, S.V.; Aben, R.W.; Renirie, R.; Wever, R.; Holtmann, D.; Rutjes, F.P.; Hollmann, F. A biocatalytic *aza*-Achmatowicz reaction. *ACS Catal.* **2016**, *6*, 5904–5907. [CrossRef]
79. But, A.; van Noord, A.; Poletto, F.; Sanders, J.P.M.; Franssen, M.C.R.; Scott, E.L. Enzymatic halogenation and oxidation using an alcohol oxidase-vanadium chloroperoxidase cascade. *Mol. Catal.* **2017**, *443*, 92–100. [CrossRef]
80. Coupe, E.E.; Smyth, M.G.; Fosberry, A.P.; Hall, R.M.; Littlechild, J.A. The dodecameric vanadium-dependent haloperoxidase from the marine algae *Corallina officinalis*: Cloning, expression, and refolding of the recombinant enzyme. *Protein Expr. Purif.* **2007**, *52*, 265–272. [CrossRef]
81. Littlechild, J. Haloperoxidases and their role in biotransformation reactions. *Curr. Opin. Chem. Biol.* **1999**, *3*, 28–34. [CrossRef]
82. Andersson, M.; Willetts, A.; Allenmark, S. Asymmetric sulfoxidation catalyzed by a vanadium-containing bromoperoxidase. *J. Org. Chem.* **1997**, *62*, 8455–8458. [CrossRef]
83. Carter, J.N.; Beatty, K.E.; Simpson, M.T.; Butler, A. Reactivity of recombinant and mutant vanadium bromoperoxidase from the red alga *Corallina officinalis*. *J. Inorg. Biochem.* **2002**, *91*, 59–69. [CrossRef]
84. Van Peè, K.-H. Microbial biosynthesis of halometabolites. *Arch. Microbiol.* **2001**, *175*, 250–258. [CrossRef] [PubMed]
85. Hofmann, B.; Tolzer, S.; Pelletier, I.; Altenbuchner, J.; van Pee, K.H.; Hecht, H.J. Structural investigation of the cofactor-free chloroperoxidases. *J. Mol. Biol.* **1998**, *279*, 889–900. [CrossRef]
86. van Peé, K.H.; Dong, C.; Flecks, S.; Naismith, J.; Patallo, E.P.; Wage, T. Biological halogenation has moved far beyond haloperoxidases. *Adv. Appl. Microbiol.* **2006**, *59*, 127–157.
87. Medici, R.; Garaycoechea, J.I.; Dettorre, L.A.; Iribarren, A.M.; Lewkowicz, E.S. Biocatalysed halogenation of nucleobase analogues. *Biotechnol. Lett.* **2011**, *33*, 1999–2003. [CrossRef]
88. Mascotti, M.L.; Ayub, M.J.; Furnham, N.; Thornton, J.M.; Laskowski, R.A. Chopping and changing: The evolution of the flavin-dependent monooxygenases. *J. Mol. Biol.* **2016**, *428*, 3131–3146. [CrossRef]
89. Dong, C.; Flecks, S.; Unversucht, S.; Haupt, C.; Van Pee, K.-H.; Naismith, J.H. Tryptophan 7-halogenase (PrnA) structure suggests a mechanism for regioselective chlorination. *Science* **2005**, *309*, 2216–2219. [CrossRef]
90. Hohaus, K.; Altmann, A.; Burd, W.; Fischer, I.; Hammer, P.E.; Hill, D.S.; Ligon, J.M.; van Pée, K.H. NADH-Dependent Halogenases Are More Likely To Be Involved in Halometaolite Biosynthesis Than Haloperoxidases. *Angew. Chem. Int. Ed.* **1997**, *36*, 2012–2013. [CrossRef]

91. Keller, S.; Wage, T.; Hohaus, K.; Hölzer, M.; Eichhorn, E.; van Pée, K.H. Purification and Partial Characterization of Tryptophan 7-Halogenase (PrnA) from *Pseudomonas fluorescens*. *Angew. Chem. Int. Ed.* **2000**, *112*, 2380–2382. [CrossRef]
92. Yeh, E.; Garneau, S.; Walsh, C.T. Robust in vitro activity of RebF and RebH, a two-component reductase/halogenase, generating 7-chlorotryptophan during rebeccamycin biosynthesis. *Proc. Natl. Acad. Sci. USA* **2005**, *102*, 3960–3965. [CrossRef]
93. Lee, J.K.; Zhao, H. Identification and characterization of the flavin:NADH reductase (PrnF) involved in a novel two-component arylamine oxygenase. *J. Bacteriol.* **2007**, *189*, 8556–8563. [CrossRef] [PubMed]
94. Eschenbrenner, M.; Covès, J.; Fontecave, M. The flavin reductase activity of the flavoprotein component of sulfite reductase from *Escherichia coli* a new model for the protein structure. *J. Biol. Chem.* **1995**, *270*, 20550–20555. [CrossRef] [PubMed]
95. Heemstra, J.R., Jr.; Walsh, C.T. Tandem action of the O2-and FADH2-dependent halogenases KtzQ and KtzR produce 6,7-dichlorotryptophan for kutzneride assembly. *J. Am. Chem. Soc.* **2008**, *130*, 14024–14025. [CrossRef]
96. Shepherd, S.A.; Karthikeyan, C.; Latham, J.; Struck, A.-W.; Thompson, M.L.; Menon, B.R.; Styles, M.Q.; Levy, C.; Leys, D.; Micklefield, J. Extending the biocatalytic scope of regiocomplementary flavin-dependent halogenase enzymes. *Chem. Sci.* **2015**, *6*, 3454–3460. [CrossRef]
97. Unversucht, S.; Hollmann, F.; Schmid, A.; van Pée, K.H. FADH2-dependence of tryptophan 7-halogenase. *Adv. Synth. Catal.* **2005**, *347*, 1163–1167. [CrossRef]
98. Schroeder, L.; Frese, M.; Müller, C.; Sewald, N.; Kottke, T. Photochemically Driven Biocatalysis of Halogenases for the Green Production of Chlorinated Compounds. *ChemCatChem* **2018**, *10*, 3336–3341. [CrossRef]
99. Ortega, M.A.; Cogan, D.P.; Mukherjee, S.; Garg, N.; Li, B.; Thibodeaux, G.N.; Maffioli, S.I.; Donadio, S.; Sosio, M.; Escano, J.; et al. Two flavoenzymes catalyze the post-translational generation of 5-chlorotryptophan and 2-aminovinyl-cysteine during NAI-107 biosynthesis. *ACS Chem. Biol.* **2017**, *12*, 548–557. [CrossRef]
100. Lang, A.; Polnick, S.; Nicke, T.; William, P.; Patallo, E.P.; Naismith, J.H.; van Pee, K.H. Changing the Regioselectivity of the Tryptophan 7-Halogenase PrnA by Site-Directed Mutagenesis. *Angew. Chem. Int. Ed.* **2011**, *50*, 2951–2953. [CrossRef]
101. Zeng, J.; Zhan, J. Characterization of a tryptophan 6-halogenase from *Streptomyces toxytricini*. *Biotechnol. Lett.* **2011**, *33*, 1607–1613. [CrossRef]
102. Hölzer, M.; Burd, W.; Reißig, H.U.; Pée, K.H.V. Substrate specificity and regioselectivity of tryptophan 7-halogenase from *Pseudomonas fluorescens* BL915. *Adv. Synth. Catal.* **2001**, *343*, 591–595. [CrossRef]
103. Hammer, P.E.; Hill, D.S.; Lam, S.T.; Van Pée, K.-H.; Ligon, J.M. Four genes from *Pseudomonas fluorescens* that encode the biosynthesis of pyrrolnitrin. *Appl. Environ. Microb.* **1997**, *63*, 2147–2154.
104. Magarvey, N.A.; Beck, Z.Q.; Golakoti, T.; Ding, Y.; Huber, U.; Hemscheidt, T.K.; Abelson, D.; Moore, R.E.; Sherman, D.H. Biosynthetic characterization and chemoenzymatic assembly of the cryptophycins. Potent anticancer agents from Nostoc cyanobionts. *ACS Chem. Biol.* **2006**, *1*, 766–779. [CrossRef] [PubMed]
105. Neumann, C.S.; Walsh, C.T.; Kay, R.R. A flavin-dependent halogenase catalyzes the chlorination step in the biosynthesis of Dictyostelium differentiation-inducing factor 1. *Proc. Natl. Acad. Sci. USA* **2010**, *107*, 5798–5803. [CrossRef] [PubMed]
106. Agarwal, V.; El Gamal, A.A.; Yamanaka, K.; Poth, D.; Kersten, R.D.; Schorn, M.; Allen, E.E.; Moore, B.S. Biosynthesis of polybrominated aromatic organic compounds by marine bacteria. *Nat. Chem. Biol.* **2014**, *10*, 640. [CrossRef]
107. Zeng, J.; Zhan, J. A Novel Fungal Flavin-Dependent Halogenase for Natural Product Biosynthesis. *ChemBioChem* **2010**, *11*, 2119–2123. [CrossRef]
108. Chinnan Velmurugan, K. Halogenases for Biosynthesis and Biocatalysis. Ph.D. Thesis, University of Manchester, Manchester, UK, 2014.
109. Andorfer, M.C.; Grob, J.E.; Hajdin, C.E.; Chael, J.R.; Siuti, P.; Lilly, J.; Tan, K.L.; Lewis, J.C. Understanding Flavin-Dependent Halogenase Reactivity via Substrate Activity Profiling. *ACS Catal.* **2017**, *7*, 1897–1904. [CrossRef]
110. Fujimori, D.G.; Hrvatin, S.; Neumann, C.S.; Strieker, M.; Marahiel, M.A.; Walsh, C.T. Cloning and characterization of the biosynthetic gene cluster for kutznerides. *Proc. Nat. Acad. Sci. USA* **2007**, *104*, 16498–16503. [CrossRef]

111. Rachid, S.; Krug, D.; Kunze, B.; Kochems, I.; Scharfe, M.; Zabriskie, T.M.; Blöcker, H.; Müller, R. Molecular and biochemical studies of chondramide formation—Highly cytotoxic natural products from *Chondromyces crocatus* Cm c5. *Chem. Biol.* **2006**, *13*, 667–681. [CrossRef]
112. Seibold, C.; Schnerr, H.; Rumpf, J.; Kunzendorf, A.; Hatscher, C.; Wage, T.; Ernyei, A.J.; Dong, C.; Naismith, J.H.; Van Pée, K.-H. A flavin-dependent tryptophan 6-halogenase and its use in modification of pyrrolnitrin biosynthesis. *Biocatal. Biotransform.* **2006**, *24*, 401–408. [CrossRef]
113. Foulston, L.C.; Bibb, M.J. Microbisporicin gene cluster reveals unusual features of lantibiotic biosynthesis in actinomycetes. *Proc. Nat. Acad. Sci. USA* **2010**, *107*, 13461–13466. [CrossRef]
114. Zehner, S.; Kotzsch, A.; Bister, B.; Süssmuth, R.D.; Méndez, C.; Salas, J.A.; van Peé, K.-H. A regioselective tryptophan 5-halogenase is involved in pyrroindomycin biosynthesis in *Streptomyces rugosporus* LL-42D005. *Chem. Biol.* **2005**, *12*, 445–452. [CrossRef]
115. Ismail, M.; Frese, M.; Patschkowski, T.; Ortseifen, V.; Niehaus, K.; Sewald, N. Flavin-Dependent Halogenases from *Xanthomonas campestris pv. campestris* B100 Prefer Bromination over Chlorination. *Adv. Synth. Catal.* **2019**, *361*, 2475–2486. [CrossRef]
116. Neubauer, P.R.; Widmann, C.; Wibberg, D.; Schroder, L.; Frese, M.; Kottke, T.; Kalinowski, J.; Niemann, H.H.; Sewald, N. A flavin-dependent halogenase from metagenomic analysis prefers bromination over chlorination. *PLoS ONE* **2018**, *13*, e0196797. [CrossRef]
117. Kirner, S.; Hammer, P.E.; Hill, D.S.; Altmann, A.; Fischer, I.; Weislo, L.J.; Lanahan, M.; van Pée, K.-H.; Ligon, J.M. Functions encoded by pyrrolnitrin biosynthetic genes from *Pseudomonas fluorescens*. *J. Bacteriol.* **1998**, *180*, 1939–1943.
118. Dairi, T.; Nakano, T.; Aisaka, K.; Katsumata, R.; Hasegawa, M. Cloning and nucleotide sequence of the gene responsible for chlorination of tetracycline. *Biosci. Biotechnol. Biochem.* **1995**, *59*, 1099–1106. [CrossRef]
119. Dorrestein, P.C.; Yeh, E.; Garneau-Tsodikova, S.; Kelleher, N.L.; Walsh, C.T. Dichlorination of a pyrrolyl-S-carrier protein by FADH2-dependent halogenase PltA during pyoluteorin biosynthesis. *Proc. Natl. Acad. Sci. USA* **2005**, *102*, 13843–13848. [CrossRef]
120. Zhang, X.; Parry, R.J. Cloning and characterization of the pyrrolomycin biosynthetic gene clusters from *Actinosporangium vitaminophilum* ATCC 31673 and *Streptomyces sp.* strain UC 11065. *Antimicrob. Agents Chemother.* **2007**, *51*, 946–957. [CrossRef]
121. Qiao, Y.; Yan, J.; Jia, J.; Xue, J.; Qu, X.; Hu, Y.; Deng, Z.; Bi, H.; Zhu, D. Characterization of the Biosynthetic Gene Cluster for the Antibiotic Armeniaspirols in *Streptomyces armeniacus*. *J. Nat. Prod.* **2019**, *82*, 318–323. [CrossRef]
122. Bayer, K.; Scheuermayer, M.; Fieseler, L.; Hentschel, U. Genomic mining for novel FADH 2-dependent halogenases in marine sponge-associated microbial consortia. *Mar. Biotechnol.* **2013**, *15*, 63–72. [CrossRef]
123. Pelzer, S.; Süssmuth, R.; Heckmann, D.; Recktenwald, J.; Huber, P.; Jung, G.; Wohlleben, W. Identification and analysis of the balhimycin biosynthetic gene cluster and its use for manipulating glycopeptide biosynthesis in *Amycolatopsis mediterranei* DSM5908. *Antimicrob. Agents Chemother.* **1999**, *43*, 1565–1573. [CrossRef]
124. Lin, S.; Van Lanen, S.G.; Shen, B. Regiospecific chlorination of (*S*)-β-tyrosyl-*S*-carrier protein catalyzed by SgcC3 in the biosynthesis of the enediyne antitumor antibiotic C-1027. *J. Am. Chem. Soc.* **2007**, *129*, 12432–12438. [CrossRef] [PubMed]
125. Wynands, I.; van Pee, K.H. A novel halogenase gene from the pentachloropseudilin producer Actinoplanes sp. ATCC 33002 and detection of in vitro halogenase activity. *FEMS Microbiol. Lett.* **2004**, *237*, 363–367. [CrossRef] [PubMed]
126. Yan, Q.; Philmus, B.; Chang, J.H.; Loper, J.E. Novel mechanism of metabolic co-regulation coordinates the biosynthesis of secondary metabolites in *Pseudomonas protegens*. *eLife* **2017**, *6*, e22835. [CrossRef] [PubMed]
127. El Gamal, A.; Agarwal, V.; Diethelm, S.; Rahman, I.; Schorn, M.A.; Sneed, J.M.; Louie, G.V.; Whalen, K.E.; Mincer, T.J.; Noel, J.P. Biosynthesis of coral settlement cue tetrabromopyrrole in marine bacteria by a uniquely adapted brominase–thioesterase enzyme pair. *Proc. Nat. Acad. Sci. USA* **2016**, *113*, 3797–3802. [CrossRef] [PubMed]
128. Podzelinska, K.; Latimer, R.; Bhattacharya, A.; Vining, L.C.; Zechel, D.L.; Jia, Z. Chloramphenicol biosynthesis: The structure of CmlS, a flavin-dependent halogenase showing a covalent flavin–aspartate bond. *J. Mol. Biol.* **2010**, *397*, 316–331. [CrossRef] [PubMed]
129. Buedenbender, S.; Rachid, S.; Müller, R.; Schulz, G.E. Structure and action of the myxobacterial chondrochloren halogenase CndH: A new variant of FAD-dependent halogenases. *J. Mol. Biol.* **2009**, *385*, 520–530. [CrossRef]

130. Menon, B.R.K.; Brandenburger, E.; Sharif, H.H.; Klemstein, U.; Shepherd, S.A.; Greaney, M.F.; Micklefield, J. RadH: A Versatile Halogenase for Integration into Synthetic Pathways. *Angew. Chem. Int. Ed.* **2017**, *56*, 11841–11845. [CrossRef]
131. Zhou, H.; Qiao, K.; Gao, Z.; Vederas, J.C.; Tang, Y. Insights into radicicol biosynthesis via heterologous synthesis of intermediates and analogs. *J. Biol. Chem.* **2010**, *285*, 41412–41421. [CrossRef]
132. Xiao, Y.; Li, S.; Niu, S.; Ma, L.; Zhang, G.; Zhang, H.; Zhang, G.; Ju, J.; Zhang, C. Characterization of tiacumicin B biosynthetic gene cluster affording diversified tiacumicin analogues and revealing a tailoring dihalogenase. *J. Am. Chem. Soc.* **2010**, *133*, 1092–1105. [CrossRef]
133. Wagner, C.; El Omari, M.; König, G.M. Biohalogenation: Nature's way to synthesize halogenated metabolites. *J. Nat. Prod.* **2009**, *72*, 540–553. [CrossRef]
134. Yeh, E.; Blasiak, L.C.; Koglin, A.; Drennan, C.L.; Walsh, C.T. Chlorination by a long-lived intermediate in the mechanism of flavin-dependent halogenases. *Biochemistry* **2007**, *46*, 1284–1292. [CrossRef] [PubMed]
135. Gkotsi, D.; Ludewig, H.; Sharma, S.; Connolly, J.; Dhaliwal, J.; Wang, Y.; Unsworth, W.; Taylor, R.; McLachlan, M.; Shanahan, S. A marine viral halogenase that iodinates diverse substrates. *Nat. Chem.* **2019**. [CrossRef] [PubMed]
136. Cahn, J.K.B.; Wilson, M.C.; Leopold-Messer, S.; Piel, J. A novel class of flavin-dependent halogenase from an uncultivated microbe suggests convergent evolution. In Proceedings of the BIOTRANS 14th International symposium on Biocatalysis and Biotransformations, Poster Abstract, Groningen, The Netherlands, 7–11 July 2019; p. 132.
137. Kemker, I.; Schnepel, C.; Schröder, D.C.C.; Marion, A.; Sewald, N. Cyclization of RGD peptides by Suzuki-Miyaura cross-coupling. *J. Med. Chem.* **2019**, *62*, 7417–7430. [CrossRef] [PubMed]
138. Mantri, M.; Krojer, T.; Bagg, E.A.; Webby, C.J.; Butler, D.S.; Kochan, G.; Kavanagh, K.L.; Oppermann, U.; McDonough, M.A.; Schofield, C.J. Crystal structure of the 2-oxoglutarate-and Fe (II)-dependent lysyl hydroxylase JMJD6. *J. Mol. Biol.* **2010**, *401*, 211–222. [CrossRef]
139. Blasiak, L.C.; Vaillancourt, F.H.; Walsh, C.T.; Drennan, C.L. Crystal structure of the non-haem iron halogenase SyrB2 in syringomycin biosynthesis. *Nature* **2006**, *440*, 368. [CrossRef]
140. Zhang, Z.; Ren, J.-S.; Harlos, K.; McKinnon, C.H.; Clifton, I.J.; Schofield, C.J. Crystal structure of a clavaminate synthase–Fe (II)–2-oxoglutarate–substrate–NO complex: Evidence for metal centred rearrangements. *FEBS Lett.* **2002**, *517*, 7–12. [CrossRef]
141. Martinez, S.; Fellner, M.; Herr, C.Q.; Ritchie, A.; Hu, J.; Hausinger, R.P. Structures and mechanisms of the non-heme Fe (II)-and 2-oxoglutarate-dependent ethylene-forming enzyme: Substrate binding creates a twist. *J. Am. Chem. Soc.* **2017**, *139*, 11980–11988. [CrossRef]
142. Wong, S.D.; Srnec, M.; Matthews, M.L.; Liu, L.V.; Kwak, Y.; Park, K.; Bell III, C.B.; Alp, E.E.; Zhao, J.; Yoda, Y. Elucidation of the Fe (IV)= O intermediate in the catalytic cycle of the halogenase SyrB2. *Nature* **2013**, *499*, 320. [CrossRef]
143. Matthews, M.L.; Neumann, C.S.; Miles, L.A.; Grove, T.L.; Booker, S.J.; Krebs, C.; Walsh, C.T.; Bollinger, J.M. Substrate positioning controls the partition between halogenation and hydroxylation in the aliphatic halogenase, SyrB2. *Proc. Natl. Acad. Sci. USA* **2009**, *106*, 17723–17728. [CrossRef]
144. Vaillancourt, F.H.; Yin, J.; Walsh, C.T. SyrB2 in syringomycin E biosynthesis is a nonheme FeII α-ketoglutarate-and O2-dependent halogenase. *Proc. Nat. Acad. Sci. USA* **2005**, *102*, 10111–10116. [CrossRef]
145. Wong, C.; Fujimori, D.G.; Walsh, C.T.; Drennan, C.L. Structural analysis of an open active site conformation of nonheme iron halogenase CytC3. *J. Am. Chem. Soc.* **2009**, *131*, 4872–4879. [CrossRef] [PubMed]
146. Ueki, M.; Galonić, D.P.; Vaillancourt, F.H.; Garneau-Tsodikova, S.; Yeh, E.; Vosburg, D.A.; Schroeder, F.C.; Osada, H.; Walsh, C.T. Enzymatic generation of the antimetabolite γ, γ-dichloroaminobutyrate by NRPS and mononuclear iron halogenase action in a streptomycete. *Chem. Biol.* **2006**, *13*, 1183–1191. [CrossRef] [PubMed]
147. Mitchell, A.J.; Zhu, Q.; Maggiolo, A.O.; Ananth, N.R.; Hillwig, M.L.; Liu, X.; Boal, A.K. Structural basis for halogenation by iron-and 2-oxo-glutarate-dependent enzyme WelO5. *Nat. Chem. Biol.* **2016**, *12*, 636. [CrossRef] [PubMed]
148. Hillwig, M.L.; Liu, X. A new family of iron-dependent halogenases acts on freestanding substrates. *Nat. Chem. Biol.* **2014**, *10*, 921–923. [CrossRef] [PubMed]

149. Hillwig, M.L.; Zhu, Q.; Ittiamornkul, K.; Liu, X. Discovery of a Promiscuous Non-Heme Iron Halogenase in Ambiguine Alkaloid Biogenesis: Implication for an Evolvable Enzyme Family for Late-Stage Halogenation of Aliphatic Carbons in Small Molecules. *Angew. Chem. Int. Ed.* **2016**, *55*, 5780–5784. [CrossRef]
150. Liu, X. In vitro Analysis of Cyanobacterial Nonheme Iron-Dependent Aliphatic Halogenases WelO5 and AmbO5. In *Methods Enzymol*; Elsevier: Amsterdam, The Netherlands, 2018; Volume 604, pp. 389–404.
151. Zhu, Q.; Liu, X. Characterization of non-heme iron aliphatic halogenase WelO5 * from *Hapalosiphon welwitschii* IC-52-3: Identification of a minimal protein sequence motif that confers enzymatic chlorination specificity in the biosynthesis of welwitindolelinones. *Beilstein J. Org. Chem.* **2017**, *13*, 1168–1173. [CrossRef]
152. Hayashi, T.; Ligibel, M.; Sager, E.; Voss, M.; Hunziker, J.; Schroer, K.; Snajdrova, R.; Buller, R. Evolved Aliphatic Halogenases Enable Regiocomplementary C-H Functionalization of an Added-Value Chemical. *Angew. Chem. Int. Ed.* **2019**, (in press). [CrossRef]
153. Flatt, P.M.; O'Connell, S.J.; McPhail, K.L.; Zeller, G.; Willis, C.L.; Sherman, D.H.; Gerwick, W.H. Characterization of the initial enzymatic steps of barbamide biosynthesis. *J. Nat. Prod.* **2006**, *69*, 938–944. [CrossRef]
154. Moosmann, P.; Ueoka, R.; Gugger, M.; Piel, J. Aranazoles: Extensively Chlorinated Nonribosomal Peptide–Polyketide Hybrids from the Cyanobacterium *Fischerella sp.* PCC 9339. *Org. Lett.* **2018**, *20*, 5238–5241. [CrossRef]
155. Matthews, M.L.; Krest, C.M.; Barr, E.W.; Vaillancourt, F.H.; Walsh, C.T.; Green, M.T.; Krebs, C.; Bollinger, J.M., Jr. Substrate-triggered formation and remarkable stability of the C− H bond-cleaving chloroferryl intermediate in the aliphatic halogenase, SyrB2. *Biochemistry* **2009**, *48*, 4331–4343. [CrossRef]
156. Welch, K.D.; Davis, T.Z.; Aust, S.D. Iron autoxidation and free radical generation: Effects of buffers, ligands, and chelators. *Arch. Biochem. Biophys.* **2002**, *397*, 360–369. [CrossRef] [PubMed]
157. Matthews, M.L.; Chang, W.-c.; Layne, A.P.; Miles, L.A.; Krebs, C.; Bollinger, J.M., Jr. Direct nitration and azidation of aliphatic carbons by an iron-dependent halogenase. *Nat. Chem. Biol.* **2014**, *10*, 209–215. [CrossRef] [PubMed]
158. Zhu, Q.; Hillwig, M.L.; Doi, Y.; Liu, X. Aliphatic Halogenase Enables Late-Stage C− H Functionalization: Selective Synthesis of a Brominated Fischerindole Alkaloid with Enhanced Antibacterial Activity. *ChemBioChem* **2016**, *17*, 466–470. [CrossRef] [PubMed]
159. Mitchell, A.J.; Dunham, N.P.; Bergman, J.A.; Wang, B.; Zhu, Q.; Chang, W.-c.; Liu, X.; Boal, A.K. Structure-guided reprogramming of a hydroxylase to halogenate its small molecule substrate. *Biochemistry* **2017**, *56*, 441–444. [CrossRef]
160. Khare, D.; Wang, B.; Gu, L.; Razelun, J.; Sherman, D.H.; Gerwick, W.H.; Håkansson, K.; Smith, J.L. Conformational switch triggered by α-ketoglutarate in a halogenase of curacin A biosynthesis. *Proc. Nat. Acad. Sci. USA* **2010**, *107*, 14099–14104. [CrossRef]
161. Vaillancourt, F.H.; Yeh, E.; Vosburg, D.A.; O'Connor, S.E.; Walsh, C.T. Cryptic chlorination by a non-haem iron enzyme during cyclopropyl amino acid biosynthesis. *Nature* **2005**, *436*, 1191. [CrossRef]
162. Schaffrath, C.; Deng, H.; O'Hagan, D. Isolation and characterisation of 5′-fluorodeoxyadenosine synthase, a fluorination enzyme from *Streptomyces cattleya*. *FEBS Lett.* **2003**, *547*, 111–114. [CrossRef]
163. O'Hagan, D.; Deng, H. Enzymatic fluorination and biotechnological developments of the fluorinase. *Chem. Rev.* **2014**, *115*, 634–649. [CrossRef]
164. Cadicamo, C.D.; Courtieu, J.; Deng, H.; Meddour, A.; O'Hagan, D. Enzymatic fluorination in *Streptomyces cattleya* takes place with an inversion of configuration consistent with an SN2 reaction mechanism. *ChemBioChem* **2004**, *5*, 685–690. [CrossRef]
165. Dong, C.; Huang, F.; Deng, H.; Schaffrath, C.; Spencer, J.B.; O'Hagan, D.; Naismith, J.H. Crystal structure and mechanism of a bacterial fluorinating enzyme. *Nature* **2004**, *427*, 561–565. [CrossRef]
166. Lohman, D.C.; Edwards, D.R.; Wolfenden, R. Catalysis by desolvation: The catalytic prowess of SAM-dependent halide-alkylating enzymes. *J. Am. Chem. Soc.* **2013**, *135*, 14473–14475. [CrossRef] [PubMed]
167. Eustáquio, A.S.; O'Hagan, D.; Moore, B.S. Engineering fluorometabolite production: Fluorinase expression in *Salinispora tropica* yields fluorosalinosporamide. *J. Nat. Prod.* **2010**, *73*, 378–382. [CrossRef] [PubMed]
168. Deng, H.; Cobb, S.L.; McEwan, A.R.; McGlinchey, R.P.; Naismith, J.H.; O'Hagan, D.; Robinson, D.A.; Spencer, J.B. The fluorinase from *Streptomyces cattleya* is also a chlorinase. *Angew. Chem. Int. Ed.* **2006**, *45*, 759–762. [CrossRef] [PubMed]

169. Lowe, P.T.; Dall'Angelo, S.; Mulder-Krieger, T.; IJzerman, A.P.; Zanda, M.; O'Hagan, D. A New Class of Fluorinated A2A Adenosine Receptor Agonist with Application to Last-Step Enzymatic [18F] Fluorination for PET Imaging. *ChemBioChem* **2017**, *18*, 2156–2164. [CrossRef] [PubMed]
170. Lowe, P.T.; Dall'Angelo, S.; Fleming, I.N.; Piras, M.; Zanda, M.; O'Hagan, D. Enzymatic radiosynthesis of a 18 F-Glu-Ureido-Lys ligand for the prostate-specific membrane antigen (PSMA). *Org. Biomol. Chem.* **2019**, *17*, 1480–1486. [CrossRef]
171. Lowe, P.T.; Dall'Angelo, S.; Devine, A.; Zanda, M.; O'Hagan, D. Enzymatic Fluorination of Biotin and Tetrazine Conjugates for Pretargeting Approaches to Positron Emission Tomography Imaging. *ChemBioChem* **2018**, *19*, 1969–1978. [CrossRef]
172. Deng, H.; Ma, L.; Bandaranayaka, N.; Qin, Z.; Mann, G.; Kyeremeh, K.; Yu, Y.; Shepherd, T.; Naismith, J.H.; O'Hagan, D. Identification of fluorinases from *Streptomyces sp.* MA37, *Norcardia brasiliensis*, and *Actinoplanes sp.* N902-109 by Genome Mining. *ChemBioChem* **2014**, *15*, 364–368. [CrossRef]
173. Ma, L.; Li, Y.; Meng, L.; Deng, H.; Li, Y.; Zhang, Q.; Diao, A. Biological fluorination from the sea: Discovery of a SAM-dependent nucleophilic fluorinating enzyme from the marine-derived bacterium *Streptomyces xinghaiensis* NRRL B24674. *RSC Adv.* **2016**, *6*, 27047–27051. [CrossRef]
174. Sun, H.; Yeo, W.L.; Lim, Y.H.; Chew, X.; Smith, D.J.; Xue, B.; Chan, K.P.; Robinson, R.C.; Robins, E.G.; Zhao, H. Directed Evolution of a Fluorinase for Improved Fluorination Efficiency with a Non-native Substrate. *Angew. Chem.* **2016**, *128*, 14489–14492. [CrossRef]
175. Yeo, W.; Chew, X.; Smith, D.; Chan, K.; Sun, H.; Zhao, H.; Lim, Y.; Ang, E. Probing the molecular determinants of fluorinase specificity. *Chem. Commun.* **2017**, *53*, 2559–2562. [CrossRef]
176. Thompson, S.; McMahon, S.A.; Naismith, J.H.; O'Hagan, D. Exploration of a potential difluoromethyl-nucleoside substrate with the fluorinase enzyme. *Bioorg. Chem.* **2016**, *64*, 37–41. [CrossRef] [PubMed]
177. Sun, H.; Zhao, H.; Ang, E. A coupled chlorinase–fluorinase system with a high efficiency of trans-halogenation and a shared substrate tolerance. *Chem. Commun.* **2018**, *54*, 9458–9461. [CrossRef] [PubMed]

Article

Mapping the Biotransformation of Coumarins through Filamentous Fungi

Jainara Santos do Nascimento [1], Wilson Elias Rozo Núñez [1], Valmore Henrique Pereira dos Santos [1], Josefina Aleu [2], Sílvio Cunha [1] and Eliane de Oliveira Silva [1,*]

1 Organic Chemistry Department, Chemistry Institute, Federal University of Bahia, Salvador 40170-115, Bahia, Brazil; naranascimento14@gmail.com (J.S.d.N.); wrozo2502@yahoo.com (W.E.R.N.); hp20.2014@gmail.com (V.H.P.d.S.); silviodc@ufba.br (S.C.)

2 Organic Chemistry Department, Faculty of Sciences, University of Cádiz, 11510 Puerto Real, Cádiz, Spain; josefina.aleu@uca.es

* Correspondence: elianeos@ufba.br; Tel.: +55-71-3283-6893

Received: 11 September 2019; Accepted: 27 September 2019; Published: 29 September 2019

Abstract: Natural coumarins are present in remarkable amounts as secondary metabolites in edible and medicinal plants, where they display interesting bioactivities. Considering the wide enzymatic arsenal of filamentous fungi, studies on the biotransformation of coumarins using these microorganisms have great importance in green chemical derivatization. Several reports on the biotransformation of coumarins using fungi have highlighted the achievement of chemical analogs with high selectivity by using mild and ecofriendly conditions. Prompted by the enormous pharmacological, alimentary, and chemical interest in coumarin-like compounds, this study evaluated the biotransformation of nine coumarin scaffolds using *Cunninghamella elegans* ATCC 10028b and *Aspergillus brasiliensis* ATCC 16404. The chemical reactions which were catalyzed by the microorganisms were highly selective. Among the nine studied coumarins, only two of them were biotransformed. One of the coumarins, 7-hydroxy-2,3-dihydrocyclopenta[*c*]chromen-4(1*H*)-one, was biotransformed into the new 7,9-dihydroxy-2,3-dihydrocyclopenta[*c*]chromen-4(1*H*)-one, which was generated by selective hydroxylation in an unactivated carbon. Our results highlight some chemical features of coumarin cores that are important to biotransformation using filamentous fungi.

Keywords: coumarin; biotransformation; filamentous fungi; selective hydroxylation

1. Introduction

Natural heterocyclic products consisting of fused benzene and α-pyrone rings are designed as coumarins [1]. The great structural diversity of coumarins can be found both in simple coumarins, whose chemical structures contain only two rings, and in coumarins that contain an additional ring, such as furano- and pyranocoumarins. In all coumarin scaffolds, hydroxy or methoxy groups at position 7 are common structural features [2].

Natural coumarins are widespread in edible and medicinal plants as secondary metabolites [3]. These chemicals display broad biological activities [4], including antioxidant [5], antibacterial [6], antiviral [7], anti-inflammatory [8], antidepressant [9], and antitumoral activities [10], among others. In addition to other secondary metabolites, the biosynthesis of coumarins is controlled by several enzymes, and their accumulation is directly influenced by biotic and abiotic factors [11]. Given the wide pharmaceutical, chemical, and alimentary importance of coumarins, several alternative ways to achieve them have been developed in an attempt to replace the natural compound obtained by bioprospecting [12,13]. As a result, limitations associated with obtaining such substances in low yields from natural sources may be circumvented.

Since the first synthetic coumarin [14], coumarin-like compounds have been mainly used in the pharmaceutical industry as precursors for the synthesis of anticoagulant drugs or in the production of fragrances. Several chemical compounds bearing a coumarin moiety continue to be produced by using synthetic methodologies [15,16], but some specific structural modifications and substituents insertions are barely performed. Sometimes, the synthetic methods require corrosive catalysts and long reaction times, and they generally generate byproducts along with the desired product [17].

Chemical derivatization using microorganisms (defined as biotransformation) represents a powerful tool for obtaining derivatives for fine chemical, pharmaceutical, and agrochemical industries according to green chemistry principles [18]. Biotransformation using microorganisms may operate in mild conditions (neutral pH, room temperature, and atmospheric pressure), and generally, it is conducive to high regio-, stereo-, and chemoselectivity at low cost [19]. Biotransformation processes have an interesting diversity, which allows for diverse products to be obtained from a single substrate [20]. Moreover, some chemical transformations that cannot be performed through the traditional synthetic methods are readily obtained using the biotransformation approach in ecofriendly reactions [21]. Within this context, the microbiological transformation of diterpenes has been studied to achieve the specific hydroxylation of unactivated C–H bonds, which is difficult to prepare using chemical methods [22].

Microbial biotransformation can be described as a reaction or a set of simultaneous reactions in which a precursor molecule is converted, rather than a fermentation process where molecules are produced from a carbon and energy source. Biotransformation could involve the use of enzymes or whole cells, or combinations thereof. In general, whole cells are more popular than isolated enzymes in industrial biotransformation [23] because the former allows for a great quantity of catalysts in small volumes and high turnover rates of enzymes and cofactors. Besides, filamentous fungi are especially useful as industrial enzyme sources because of their broad diversity in the production of proteins, which are quickly secreted [24].

Recently, our research group published a review on the biotransformation of simple, furano-, and pyranocoumarins using different genera of filamentous fungi [25]. The survey showed that *Cunninghamella* sp. and *Aspergillus* sp. are the most common genera used as catalysts in the biotransformation of several types of coumarins. These findings prompted us to investigate the chemical specificities of *Cunninghamella* and *Aspergillus* genera in transforming coumarin cores.

Therefore, the present study reports on the biotransformation of a variety of coumarin compounds using *Cunninghamella elegans* ATCC 10028b and *Aspergillus brasiliensis* ATCC 16404. Our results provide an interesting way of mapping these microbial systems onto the derivatization of coumarin scaffolds. The approach used by us highlights some chemical features in the coumarin structures that allow for their biotransformation. Additionally, the biotransformation of coumarins using the selected fungi strains occurred under chemo- and stereoselectivity. One of the assays led to a selective hydroxylation at an unexpected position of the coumarin core.

2. Results and Discussion

Coumarins represent an important class of natural products and also synthetic oxygen-containing heterocycles. Several reports have claimed that the most common structural pattern in natural coumarins is hydroxylation at C-7. Based on this, the present study focused on an evaluation of the biotransformation of a panel of coumarins that contained one or two hydroxyl groups in the aromatic ring.

Initial screening of the biotransformation of nine coumarin compounds (**1** to **9**, Figure 1) was carried out using the filamentous fungi *Cunninghamella elegans* ATCC 10028b and *Aspergillus brasiliensis* ATCC 16404. The biotransformations were carried out for 72 h, and samples were analyzed using HPLC every 24 h. Both fungi were able to transform only two of the coumarins (**4** and **7**) after 72 h of incubation.

	R_1	R_2
1	H	H
2	OH	H
3	H	OH

	R_1	R_2
4	H	H
5	OH	H
6	H	OH

	R_1	R_2
7	H	H
8	OH	H
9	H	OH

Figure 1. Chemical structures of coumarins used as substrates in biotransformation through *Cunninghamella elegans* ATCC 10028b and *Aspergillus brasiliensis* ATCC 16404.

HPLC analysis (Figure 2) showed that *C. elegans* and *A. brasiliensis* biotransformed (exclusively) the coumarins **4** and **7** into more polar derivatives. The chemical structures of the coumarin substrates **4** and **7** contained two common features: aromatic rings from both contained only one hydroxyl group at C-7, and both coumarins contained a bulky group at C-4. None of the coumarins whose aromatic ring contained two hydroxyl groups (**2**, **3**, **5**, **6**, **8**, and **9**) were biotransformed by *C. elegans* or *A. brasiliensis*. Moreover, none of the coumarins with a methyl group at C-4 (**1**, **2**, and **3**) were biotransformed by the two evaluated filamentous fungi.

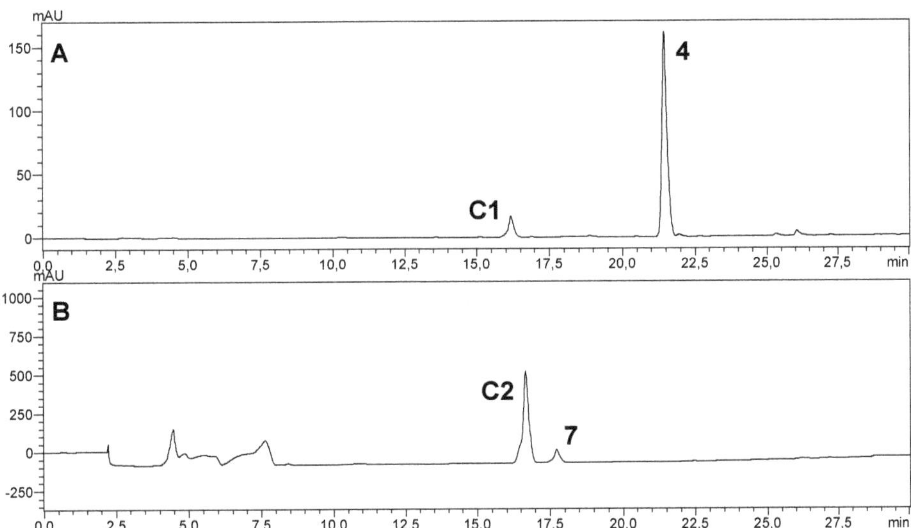

Figure 2. Reverse-phase HPLC elution profiles (λ = 211 nm) of the ethyl acetate extracts of *C. elegans* ATCC 10028b cultures incubated for 72 h with coumarin **4** (chromatogram A) and *A. brasiliensis* ATCC 16404 cultures incubated for 72 h with coumarin **7** (chromatogram B). The derivatives of coumarins **4** and **7** can be visualized in peaks **C1** and **C2**, respectively. AU: absorbance unit.

A more detailed analysis of the HPLC chemical profiles of the crude extracts of the biotransformations catalyzed by *C. elegans* and *A. brasiliensis* showed that both fungi biotransformed coumarins **4** and **7** into the same derivatives. However, the yields of derivatives **C1** and **C2** were different. Concerning the biotransformation of coumarin **4**, the yield of the assay catalyzed by *C. elegans* was five-fold higher than that catalyzed by *A. brasiliensis*. On the other hand, *A. brasiliensis* was more efficient in the biotransformation of coumarin **7**, generating almost double the yield. Similarly to our study, Lee and coworkers reported the biotransformation of isoflavone by *C. elegans* and *A. niger* [26]. The authors highlighted that *A. niger* gave a more complex metabolite profile than did *C. elegans*, but some derivatives were found in the crude extracts from both fungi.

Next, the biotransformation experiments were repeated to facilitate the isolation and characterization of **C1** and **C2**. Scale-up biotransformations of coumarin-related compounds **4** and **7** by *C. elegans* and *A. brasiliensis*, respectively, led to the isolation of their main derivatives.

The derivative isolated from the biotransformation of coumarin **4** was identified as 7,9-dihydroxy-2,3-dihydrocyclopenta[*c*]chromen-4(1*H*)-one (**C1**), which is reported for the first time in the present study. Its molecular formula was determined as $C_{12}H_{10}O_4$ by HRESIMS (*m/z* 217.0507 $[M-H]^-$, Figure S6 in the Supplementary Materials), indicating that one new hydroxyl group was introduced into **4** as a result of the biotransformation. A comparison between the **4** and **C1** ^{1}H NMR spectra (Figure S1 in the Supplementary Materials) showed that their chemical structures were very similar, except for C-9 (δ 75.1) in the **C1** chemical structure. The chemical shift of C-9 (δ 75.1) and the presence of only one hydrogen attached at C-9 (δ 5.20–5.22, ddd, H-9) indicated hydroxylation at this position. Heteronuclear HSQC and HMBC analysis, along with ^{13}C NMR data analysis (see spectra in Figures S2–S4 in the Supplementary Materials), allowed for the unequivocal structural identification of **C1** (Figure 3). The substitution of one hydrogen for the hydroxyl group at C-9 explained the greater polarity of **C1** compared to its precursor, **4** (see chromatogram A, Figure 2).

The hydroxylation at C-9 generated a chiral center in the **C1** chemical structure, which explained the multiplicity of the diastereotopic hydrogens attached at C-10 (δ 2.42–2.48, dddd, H-10a; and 2.00–2.05, dddd, H-10b) and C-11 (δ 3.17–3.23, dddd, H-11a; and 2.95–3.00 ddd, H-11b). The ddd δ 5.20–5.22 was attributed to the enantiotopic hydrogen attached at C-9. The optical rotation measurement of the **C1** sample ($[\alpha]^{20}{}_D = -16°$) showed that the biotransformation approach used herein was stereoselective with a preference for levogyre stereoisomer production.

The magnitude of the *J* between two adjacent C–H bonds ($^3J_{HH}$) is directly dependent on the dihedral angle α between these two bonds [27]. Therefore, the *J* analysis of the ^{1}H NMR signals of the hydrogens belonging to the **C1** cyclopentanol ring led to some conclusions about the relative stereochemistry of its substituents. The largest *J* value of ddd δ 5.20–5.22 ($^3J_{HH}$ = 6.9 Hz) was attributed to the vicinal coupling between H-9 and H-10a (δ 2.42–2.48) that should occupy the opposite face of the cyclopentanol ring. By the same rationale, the middle *J* value ($^3J_{HH}$ = 2.2 Hz) was attributed to the vicinal coupling between H-9 and H-10b (δ 2.00–2.05) that should occupy the same face of the cyclopentanol ring. Finally, the smallest *J* value of the ddd ($^4J_{HH}$ = 1.4 Hz) was attributed to the W-coupling between H-9 and H-11a (δ 3.17–3.23). The W-coupling was herein confirmed through COSY homonuclear analysis (Figure S5 in the Supplementary Materials).

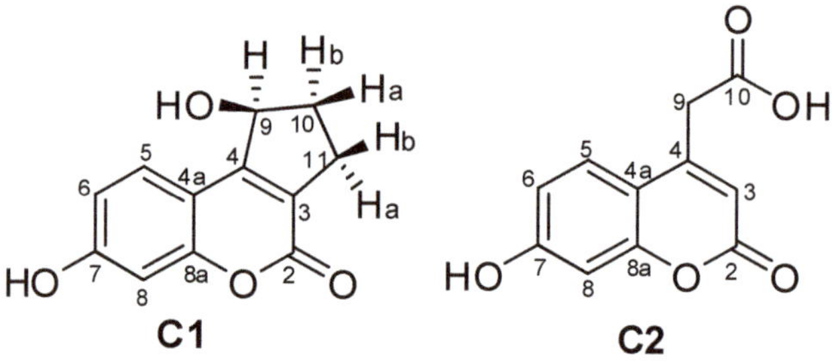

Figure 3. Chemical structures of **C1** and **C2**, which were isolated from the biotransformation of coumarins **4** and **7**, respectively.

Several coumarins and their derivatives have displayed interesting biological activities, and they are useful as a starting point for drug development [25]. The biotransformation approach employed in the present study provided chemo- and stereoselective hydroxylation of an unactivated C–H bond (position 9 of coumarin **4**). The ability of microorganisms to hydroxylate chemically inaccessible centers is a powerful synthetic tool because the functionalization of unactivated C–H bonds is a true challenge to organic synthesis [28]. Traditional chemical methods generally require highly reactive oxidizing agents, which causes difficulties in the regio- and stereocontrols.

Considering the poor conversion rate of coumarin **4** and the consequent low yield of **C1** (9.0% yield), we investigated the biotransformation of **4** by *C. elegans*, intending to improve its conversion rate. The biotransformation of coumarin **4** by *C. elegans* was then carried out by using twice the amount of fungus and also by changing the incubation time to 120 and 168 h. We concluded that the biotransformation assay that used twice the amount of fungus for 72 h was the best one for the biotransformation of **4** into **C1**. Next, we repeated the biotransformation of **4** by *C. elegans* by using the optimized conditions, and we isolated **C1** with a yield of 22.0%.

The next step of our study was the identification of the chemical structure of the derivative **C2**, which was achieved through biotransformation of the coumarin **7** by *A. brasiliensis* with a yield of 35.0%. The **C2** ^{1}H NMR spectrum analysis (Figure S7 in the Supplementary Materials) showed that the biotransformation of **7** by *A. brasiliensis* led to hydrolysis at C-10. The ester group at C-10 of the chemical structure of **7** was converted into a carboxyl acid group in **C2**. The **C2** and **7** ^{1}H NMR spectra were almost identical. Meanwhile, no methyl hydrogen was present in the **C2** ^{1}H NMR spectrum. The ^{1}H NMR of coumarin **7** contained a singlet at δ 3.83 with an integral value of 3. This signal was not found in the **C2** ^{1}H NMR spectrum. All **C2** NMR data were in accord with those previously reported for 2-(7-hydroxy-2-oxo-2*H*-chromen-4-yl) acetic acid or 7-hydroxycoumarinyl-4-acetic acid [29].

It was previously established that C-7 substitution in the coumarin nucleus increases its antioxidant and antifungal activities [30]. Within this context, Molnar and coworkers have synthesized some coumarinyl thiosemicarbazides from **C2**. The synthesized compounds displayed interesting antibacterial activity against *Bacillus subtilis* [31].

As part of our studies on the biotransformation of coumarins, we recently reported that transformations at C-7, reductions at C3–C4, and lactone-ring opening were the most frequent reactions in coumarin cores submitted to biotransformation using filamentous fungi [25]. Unlike what was expected, the biotransformations described in the present study led to hydroxylation at an unexpected position and hydrolysis reaction.

In summary, our investigations into the biotransformation of coumarin compounds by *C. elegans* and *A. brasiliensis* provided useful information about structural features that are important for the microbial transformation of this kind of compound. Coumarins that contained aromatic monohydroxylation and a bulky group at C-4 in their chemical structures were efficiently biotransformed by the fungi strains. Although both fungi strains converted the coumarins into the same derivatives, the reaction yields were distinct. We demonstrated the occurrence of a stereoselective hydroxylation at an unactivated carbon of one of the coumarins with a preference for levogyre stereoisomer production. The spectroscopic analysis allowed for the identification of a new chiral coumarin, whose relative stereochemistry was identified.

3. Materials and Methods

3.1. General Analytical Procedures

Nuclear magnetic resonance (NMR) spectra were recorded in CD_3OD on a Varian VNMRS 600 (^{1}H: 600 MHz; ^{13}C: 150 MHz; Palo Alto, CA, USA) spectrometer operating at 25 °C or in DMSO-d_6 on a Varian NMR AS 400 (^{1}H: 400 MHz) spectrometer operating at 25 °C. The chemical shifts (δ) were assigned in ppm and the coupling constants (J) in Hz. The assignments were based on chemical shifts, integration, homonuclear (COSY), and heteronuclear (HMQC and HMBC) measurements. Optical rotation was measured at 20 °C in a Perkin Elmer 343 Polarimeter (Waltham, MA, USA) at 589 nm (sodium *D* line). Analytical HPLC analyses were carried out on a Shimadzu Shim-pack PREP-ODS(H)KIT 5 µm C_{18} column (4.6 × 250.0 mm, Kyoto, Japan), and the chemical profiles of the biotransformation and control extracts were obtained using 10% to 100% methanol in water containing 0.01% acetic acid over 30 min with a flow rate of 0.8 mL/min. The crude extracts were analyzed through the injection of 20 μL at 1 mg/mL on a Shimadzu (SIL-20A) multisolvent delivery system, a Shimadzu SPD-M20A, a photodiode array detector, and an Intel Celeron computer for analytical system control, data collection, and processing. The derivatives were isolated by using a chromatographic column (40 × 1.5 cm) containing silica gel (Sigma-Aldrich, 60 Å, Saint Louis, MO, USA). Mixtures of *n*-hexane (Synth) and ethyl acetate (Synth) were employed as a mobile phase.

3.2. Substrates

Nine synthetic coumarin analogs were submitted to the biotransformation experiments. All coumarins used as substrates were obtained according to known methods through a Pechmann reaction, and all of their spectroscopic data were identical to those previously described for 7-hydroxy-4-methyl-2*H*-2-chromenone (**1**) [32], 5,7-dihydroxy-4-methyl-2*H*-chromen- 2-one (**2**) [33], 7,8-dihydroxy-4-methyl-2*H*-chromen-2-one (**3**) [34], 7-hydroxy-2,3-dihydrocyclopenta[*c*]chromen-4(1*H*)-one (**4**) [33], 7,9-dihydroxy-2,3-dihydrocyclopenta[*c*]chromen-4(1*H*)-one (**5**) [33], 6,7-dihydroxy-2,3-dihydrocyclopenta[*c*]chromen-4(1*H*)-one (**6**) [35], methyl 2-(7,8-dihydroxy-2-oxo-2*H*-chromen-4-yl)acetate (**7**) [36], methyl 2-(5,7-dihydroxy-2-oxo-2*H*-chromen-4-yl)acetate (**8**) [37], and methyl 2-(7-hydroxy-2-oxo-2*H*-chromen-4-yl)acetate (**9**) [38].

3.3. Biotransformation Assays

The biotransformation of all coumarins (**1–9**) was done through screening with *Cunninghamella elegans* ATCC 10028b and *Aspergillus brasiliensis* ATCC 16404, which were obtained from the American Type Culture Collection (ATCC, Rockville, MD, USA). The filamentous fungi were maintained in 80% glycerol solution at −20 °C.

The fungi were grown in a two-step culture procedure. First, each fungus was grown at 28 °C in Petri dishes containing malt agar (malt extract 2.0%, glucose 2.0%, peptone 0.1%, agar 1.8%) for 7 days. Next, an inoculum of five 6-mm disks containing mycelia and agar was added to 125-mL Erlenmeyer flasks, each holding 50 mL of Koch's K1 medium (glucose 0.18%, peptone 0.06%, and yeast extract 0.04%). Coumarins (5 mg) were separately added to each flask as a solution in dimethyl sulfoxide (5 mg dissolved in 200 μL). Control flasks consisted of a culture medium with dimethyl sulfoxide and a fungus (without coumarin), a culture medium with coumarin and dimethyl sulfoxide (without a fungus), and a culture medium by itself. Biotransformation experiments were carried out at 28 °C for 72 h with shaking at 120 rpm. Samples were analyzed daily by HPLC. The mycelia were separated by filtration, the fermentation broths were extracted three times with ethyl acetate, and the solvent was evaporated under reduced pressure to yield crude extracts. All experiments were carried out in triplicate.

Biotransformations of two selected coumarins (**4** and **7**) were separately carried out in 10 Erlenmeyer flasks (scale-up biotransformations) using the same aforementioned procedures. According to the yields of the biotransformation assays in the initial screening, the scale-up biotransformations of **4** and **7** were carried out by *C. elegans* and *A. brasiliensis*, respectively. The extraction of the culture broths by ethyl acetate was followed by evaporation of the solvent to yield the crude extracts of the biotransformations of **4** and **7** (41.0 mg and 59.0 mg, respectively).

Additionally, the biotransformation conditions of coumarin **4** were investigated with a view toward increasing its conversion. For this, new assays were designed by using a greater quantity of the fungus *C. elegans* (10 6-mm disks containing mycelia and agar) and by increasing the incubation time (120 and 168 h).

3.4. Isolation and Purification of the Derivatives

The extracts from the culture broths of *C. elegans* or *A. brasiliensis* (scale-up biotransformations) with coumarins **4** and **7** were submitted to an isolation procedure (as described in Section 3.1) to yield the derivatives **C1** and **C2**, respectively.

3′,7-dihydroxy-2,3-dihydrocyclopenta[*c*]chromen-4(1*H*)-one (**C1**): 3.6 mg (9.0% yield) brown powder; $[\alpha]^{20}_D$ =−16° (0.34; CH_3OH); ^1H NMR (600 MHz, CD_3OD) 7.44 (d, *J* = 8.4 Hz, 1H, H-5), 6.77 (dd, *J* = 8.4 and 2.2 Hz, 1H, H-6), 6.68 (d, *J* = 2.2 Hz, 1H, H-8), 5.20–5.22 (ddd, *J* = 6.9, 2.2, and 1.4 Hz, 1H, H-9), 3.17–3.23 (dddd, *J* = 18.0, 8.0, 6.9, and 1.4 Hz, 1H, H-11a), 2.95–3.00 (ddd, *J* = 18.0, 9.1, and 3.3 Hz, 1H, H-11b), 2.42–2.48 (dddd, *J* = 13.5, 9.1, 6.9, and 6.9 Hz, 1H, H-10a), 2.00–2.05 (dddd, *J* = 13.5, 8.0, 3.3, and 2.2 Hz, 1H, H-10b); ^{13}C NMR (150 MHz, CD_3OD) 163.1 (C-2), 162.2 (C-7), 161.3 (C-4), 158.2 (C-8a), 127.8 (C-5 and C-3), 115.5 (C-6), 103.8 (C-4a and C-8), 75.1 (C-9), 34.2 (C-10), 30.3 (C-11). HRESIMS *m/z* 217.0507 $[M -H]^-$ (calcd for $[M -H]^-$ 217.0501).

2-(7-hydroxy-2-oxo-2*H*-chromen-4-yl) acetic acid (**C2**): 13.0 mg (35.0% yield) brown powder; ^1H NMR (400 MHz, DMSO-d_6) 12.75 (br s, 1H, COOH), 10.57 (s, 1H, OH-7), 7.54 (d, *J* = 8.7 Hz, 1H, H-5), 6.82 (dd, *J* = 8.7 and 2.3 Hz, 1H, H-6), 6.74 (d, *J* = 2.3 Hz, 1H, H-8), 6.23 (s, 1H, H-3), 3.83 (s, 2H, H-9).

Supplementary Materials: The following are available online, Figure S1: ^1H NMR spectrum of compound **C1** (600 MHz, CD_3OD); Figure S2: ^{13}C NMR spectrum of compound **C1** (150 MHz, CD_3OD); Figure S3: ^1H-^{13}C HSQC 2D NMR correlation spectroscopy of compound **C1** (600 MHz/150 MHz, CD_3OD); Figure S4: ^1H-^{13}C HMBC 2D NMR correlation spectroscopy of compound **C1** (600 MHz/150 MHz, CD_3OD); Figure S5: ^1H-^1H COSY 2D NMR correlation spectroscopy of compound **C1** (600 MHz, CD_3OD); Figure S6: HRESIMS spectrum of compound **C1** (negative ion mode); Figure S7: ^1H NMR spectrum of compound **C2** (400 MHz, DMSO-d_6).

Author Contributions: J.S.d.N. and V.H.P.d.S. carried out the biotransformation experiments, W.E.R.N. and S.C. synthetized the coumarins, J.A. acquired and interpreted the NMR and MS spectra, and E.d.O.S. designed the experiments, interpreted the NMR spectra, and wrote the paper.

Funding: The authors are grateful to the "Bahia Research Foundation" (FAPESB), "Conselho Nacional de Desenvolvimento Científico e Tecnológico" (CNPq), and "Coordenacão de Aperfeiçoamento de Pessoal de Nível Superior" (CAPES, Finance Code 001).

Acknowledgments: E.O.S. thanks Niege Araçari Jacometti Cardoso Furtado for kindly giving the fungal strains.

Conflicts of Interest: The authors declare no conflicts of interest.

References

1. Matos, M.J.; Santana, L.; Uriarte, E.; Abreu, O.A.; Molina, E.; Yordi, E.G. Coumarins—An important class of phytochemicals. In *Phytochemicals—Isolation, Characterisation and Role in Human Health*; Rao, A.V., Rao, L.G., Eds.; Intech: London, UK, 2015; pp. 113–140.
2. Bone, K.; Mills, S. *Principles and Practice of Phytotherapy—Modern Herbal Medicine*, 2nd ed.; Elsevier: Atlanta, GA, USA, 2013; p. 1076.
3. Witaicenis, A.; Seito, L.N.; da Silveira Chagas, A.; de Almeida, L.D.; Luchini, A.C.; Rodrigues-Orsi, P.; Cestari, S.H.; Di Stasi, L.C. Antioxidant and intestinal anti-inflammatory effects of plant-derived coumarin derivatives. *Phytomedicine* **2014**, *21*, 240–246. [CrossRef] [PubMed]
4. Venugopala, K.N.; Rashmi, V.; Odhav, B. Review on natural coumarin lead compounds for their pharmacological activity. *Biomed Res. Int.* **2013**, *2013*, 963248. [CrossRef] [PubMed]
5. Borges Bubols, G.; da Rocha Vianna, D.; Medina-Remon, A.; von Poser, G.; Maria Lamuela-Raventos, R.; Lucia Eifler-Lima, V.; Cristina Garcia, S. The antioxidant activity of coumarins and flavonoids. *Mini Rev. Med. Chem.* **2013**, *13*, 318–334. [CrossRef] [PubMed]
6. de Souza, S.M.; Monache, F.D.; Smânia, A.J. Antibacterial activity of coumarins. *Z. Naturforsch.* **2005**, *60*, 693–700. [CrossRef] [PubMed]
7. Hassan, M.Z.; Osman, H.; Ali, M.A.; Ahsan, M.J. Therapeutic potential of coumarins as antiviral agents. *Eur. J. Med. Chem.* **2016**, *123*, 236–255. [CrossRef] [PubMed]
8. Bansal, Y.; Sethi, P.; Bansal, G. Coumarin: A potential nucleus for anti-inflammatory molecules. *Med. Chem. Res.* **2013**, *22*, 3049–3060. [CrossRef]

9. Capra, J.C.; Cunha, M.P.; Machado, D.G.; Zomkowski, A.D.E.; Mendes, B.G.; Santos, A.R.S.; Pizzolatti, M.G.; Rodrigues, A.L.S. Antidepressant-like effect of scopoletin, a coumarin isolated from *Polygala sabulosa* (Polygalaceae) in mice: Evidence for the involvement of monoaminergic systems. *Eur. J. Pharmacol.* **2010**, *643*, 232–238. [CrossRef]
10. Kawase, M.; Sakagami, H.; Motohashi, N.; Hauer, H.; Chatterjee, S.S.; Spengler, G.; Vigyikanne, A.V.; Molnár, A.; Molnár, J. Coumarin derivatives with tumor-specific cytotoxicity and multidrug resistance reversal activity. *In Vivo* **2005**, *19*, 705–712.
11. Gobbo-Neto, L.; Lopes, N.P. Plantas medicinais: Fatores de influência no conteúdo de metabólitos secundários. *Quim. Nova* **2007**, *30*, 374–381. [CrossRef]
12. Mira, A.; Alkhiary, W.; Zhu, Q.; Nakagawa, T.; Tran, H.B.; Amen, Y.M.; Shimizu, K. Improved biological activities of isoepoxypteryxin by biotransformation. *Chem. Biodivers.* **2016**, *13*, 1307–1315. [CrossRef]
13. Yang, X.; Hou, J.; Liu, D.; Deng, S.; Wang, Z.B.; Kuang, H.X.; Wang, C.; Yao, J.H.; Liu, K.X.; Ma, X.C. Biotransformation of isoimperatorin by *Cunninghamella blakesleana* AS 3.970. *J. Mol. Catal. B Enzym. B* **2013**, *88*, 1–6. [CrossRef]
14. Vargas-Soto, F.A.; Céspedes-Acuña, C.L.; Aqueveque- Muñoz, P.M.; Alarcón-Enos, J.E. Toxicity of coumarins synthesized by Pechmann-Duisberg condensation against *Drosophila melanogaster* larvae and antibacterial effects. *Food Chem. Toxicol.* **2017**, *109*, 1118–1124. [CrossRef] [PubMed]
15. Zareyee, D.; Serehneh, M. Recyclable CMK-5 supported sulfonic acid as an environmentally benign catalyst for solvent-free one-pot construction of coumarin through Pechmann condensation. *J. Mol. Catal. A Chem.* **2014**, *391*, 88–91. [CrossRef]
16. Maheswara, M.; Siddaiah, V.; Damu, G.L.V.; Rao, Y.K.; Rao, C.V. A solvent-free synthesis of coumarins via Pechmann condensation using heterogeneous catalyst. *J. Mol. Catal. A Chem.* **2006**, *255*, 49–52. [CrossRef]
17. Vekariya, R.H.; Patel, H.D. Recent advances in the synthesis of coumarin derivatives via Knoevenagel condensation: A review. *Synth. Commun. Rev.* **2014**, *44*, 2756–2788. [CrossRef]
18. Watson, W.J.W. How do the fine chemical, pharmaceutical, and related industries approach green chemistry and sustainability? *Green Chem.* **2012**, *14*, 251–259. [CrossRef]
19. Hai-Feng, Z.; Guo-Qing, H.; Jing, L.; Hui, R.; Qi-He, C.; Qiang, Z.; Jin-Ling, W.; Hong-Bo, Z. Production of gastrodin through biotransformation of *p*-2-hydroxybenzyl alcohol by cultured cells of *Armillaria luteo-virens* Sacc. *Enzyme Microb. Technol.* **2008**, *43*, 25–30. [CrossRef]
20. Müller, M. Chemical diversity through biotransformations. *Curr. Opin. Biotechnol.* **2004**, *15*, 591–598. [CrossRef]
21. Silva, E.O.; Furtado, N.A.J.C.; Aleu, J.; Collado, I.G. Non-terpenoid biotransformations by *Mucor* species. *Phytochem. Rev.* **2015**, *14*, 745–764. [CrossRef]
22. Fraga, B.M.; Gonzalez-vallejo, V.; Guillermo, R. On the biotransformation of *ent*-trachylobane to *ent*-kaur-11-ene diterpenes. *J. Nat. Prod.* **2011**, *74*, 1985–1989. [CrossRef]
23. Straathof, A.J.; Panke, S.; Schmid, A. The production of fine chemicals by biotransformations. *Curr. Opin. Biotechnol.* **2002**, *13*, 548–556. [CrossRef]
24. Corrêa, R.C.G.; Rhoden, S.A.; Mota, T.R.; Azevedo, J.L.; Pamphile, J.A.; de Souza, C.G.M.; de Moraes, M.D.L.T.; Bracht, A.; Peralta, R.M. Endophytic fungi: Expanding the arsenal of industrial enzyme producers. *J. Ind. Microbiol. Biotechnol.* **2014**, *41*, 1467–1478. [CrossRef] [PubMed]
25. Do Nascimento, J.S.; Conceição, J.C.S.; de Oliveira Silva, E. Biotransformation of coumarins by filamentous fungi: An alternative way for achievement of bioactive analogs. *Mini. Rev. Org. Chem.* **2019**, *16*, 568–577. [CrossRef]
26. Lee, J.H.; Oh, E.T.; Chun, S.C.; Keum, Y.S. Biotransformation of isoflavones by *Aspergillus niger* and *Cunninghamella elegans*. *J. Korean Soc. Appl. Biol. Chem.* **2014**, *57*, 523–527. [CrossRef]
27. Pavia, D.L.; Lampman, G.M.; Kriz, G.S.; Vyvyan, J.A. *Introduction to Spectroscopy*, 5th ed.; Cengage Learning: Boston, MA, USA, 2014; p. 784.
28. Fraga, B.M.; Guillermo, R.; Herna, M.G.; Garbarino, J.A. Biotransformation of two stemodane diterpenes by *Mucor plumbeus*. *Tetrahedron* **2004**, *60*, 7921–7932. [CrossRef]
29. Dubovik, I.P.; Garazd, M.M.; Khilya, V.P. Modified coumarins. 14. Synthesis of 7-hydroxy-[4,3′]dichromenyl-2, 2′-dione derivatives. *Chem. Nat. Compd.* **2004**, *40*, 434–443. [CrossRef]
30. Molnar, M.; Šarkanj, B.; Čačić, M.; Gille, L.; Strelec, I. Antioxidant properties and growth-inhibitory activity of coumarin Schiff bases against common foodborne fungi. *Der Pharma Chem.* **2014**, *6*, 313–320.

31. Molnar, M.; Tomić, M.; Pavić, V. Coumarinyl thiosemicarbazides as antimicrobial agents. *Pharm. Chem. J.* **2018**, *51*, 1078–1081. [CrossRef]
32. Ma, J.; Zhang, G.; Han, X.; Bao, G.; Wang, L.; Zhai, X.; Gong, P. Synthesis and biological evaluation of benzothiazole derivatives bearing the *ortho*-hydroxy-*N*-acylhydrazone moiety as potent antitumor agents. *Arch. Pharm. (Weinh.)* **2014**, *347*, 936–949. [CrossRef]
33. Prateeptongkum, S.; Duangdee, N.; Thongyoo, P. Facile iron(III) chloride hexahydrate catalyzed synthesis of coumarins. *ARKIVOC* **2015**, *2015*, 248–258. [CrossRef]
34. Shen, Q.; Shao, J.; Peng, Q.; Zhang, W.; Ma, L.; Chan, A.S.C.; Gu, L. Hydroxycoumarin derivatives: Novel and potent α-glucosidase inhibitors. *J. Med. Chem.* **2010**, *53*, 8252–8259. [CrossRef] [PubMed]
35. Liu, L.; Zhang, M.; Zhou, L.; Zhu, J.; Guan, S.; Yu, R. Biotransformation of 3,4-cyclocondensed coumarins by transgenic hairy roots of *Polygonum multiflorum*. *Afr. J. Pharm. Pharmacol.* **2012**, *6*, 3047–3054. [CrossRef]
36. Lizzul-Jurse, A.; Bailly, L.; Hubert-Roux, M.; Afonso, C.; Renard, P.Y.; Sabot, C. Readily functionalizable phosphonium-tagged fluorescent coumarins for enhanced detection of conjugates by mass spectrometry. *Org. Biomol. Chem.* **2016**, *14*, 7777–7791. [CrossRef] [PubMed]
37. Smitha, G.; Reddy, C.S. $ZrCl_4$-catalyzed Pechmann reaction: Synthesis of coumarins under solvent-free conditions. *Synth. Commun.* **2004**, *34*, 3997–4003. [CrossRef]
38. Dubovik, I.P.; Garazd, M.M.; Khilya, V.P. Modified coumarins. 14. Synthesis of 7-hydroxy-[4,3′]dichromenyl-2,2′-dione derivatives. *Chem. Nat. Compd.* **2004**, *40*, 358–365. [CrossRef]

Sample Availability: Samples of the compounds are available from the authors.

Article

A Gram-Scale Limonene Production Process with Engineered *Escherichia coli*

Jascha Rolf [1], Mattijs K. Julsing [1,2], Katrin Rosenthal [1] and Stephan Lütz [1,*]

1 Chair for Bioprocess Engineering, Department of Biochemical and Chemical Engineering, TU Dortmund University, D-44227 Dortmund, Germany; jascha.rolf@tu-dortmund.de (J.R.); mattijs.julsing@wur.nl (M.K.J.); katrin.rosenthal@tu-dortmund.de (K.R.)

2 Wageningen Food & Biobased Research, Wageningen University & Research, 6708 WG Wageningen, The Netherlands

* Correspondence: stephan.luetz@tu-dortmund.de; Tel.: +49-231-755-4764

Received: 31 March 2020; Accepted: 16 April 2020; Published: 18 April 2020

Abstract: Monoterpenes, such as the cyclic terpene limonene, are valuable and important natural products widely used in food, cosmetics, household chemicals, and pharmaceutical applications. The biotechnological production of limonene with microorganisms may complement traditional plant extraction methods. For this purpose, the bioprocess needs to be stable and ought to show high titers and space-time yields. In this study, a limonene production process was developed with metabolically engineered *Escherichia coli* at the bioreactor scale. Therefore, fed-batch fermentations in minimal medium and in the presence of a non-toxic organic phase were carried out with *E. coli* BL21 (DE3) pJBEI-6410 harboring the optimized genes for the mevalonate pathway and the limonene synthase from *Mentha spicata* on a single plasmid. The feasibility of glycerol as the sole carbon source for cell growth and limonene synthesis was examined, and it was applied in an optimized fermentation setup. Titers on a gram-scale of up to 7.3 $g \cdot L_{org}^{-1}$ (corresponding to 3.6 $g \cdot L^{-1}$ in the aqueous production phase) were achieved with industrially viable space-time yields of 0.15 $g \cdot L^{-1} \cdot h^{-1}$. These are the highest monoterpene concentrations obtained with a microorganism to date, and these findings provide the basis for the development of an economic and industrially relevant bioprocess.

Keywords: monoterpenes; limonene; glycerol; mevalonate pathway; reaction engineering; bioprocess; biocatalyst; two-liquid phase fermentation; in situ product removal

---

1. Introduction

Monoterpenes are volatile, lipophilic compounds in the essential oils of plants, which often find application as flavors and fragrances in food, cosmetics, and household chemicals. Limonene is the predominant monoterpene in the essential oils of citrus fruits and can be found in oaks, pines, and spearmint as well. Recently, limonene has been investigated as a promising alternative or additive for solvents [1] and jet fuels [2–4]. Limonene also shows antimicrobial properties [5], can be easily functionalized because of its two double bonds [6], and thus finds application as a building block for several commodity chemicals and pharmaceuticals. The oxygenated derivatives of limonene show potent pharmaceutical activities. As an example, perillyl alcohol, which can be obtained by the regiospecific oxygenation of limonene via whole-cell biotransformation [7,8], has proven anti-cancer properties [9]. The application of monoterpenes as starting materials for industrially or pharmaceutically relevant compounds requires efficient synthesis routes [10]. Nowadays, limonene is mainly produced as a by-product of orange juice production. However, the establishment of new applications will lead to a rapidly growing global market. The low concentrations of monoterpenes in natural sources make their isolation often economically unfeasible. Chemical synthesis might offer alternative production strategies. However, the chemical synthesis of these complex and often chiral molecules is typically difficult, involves many synthesis steps, and suffers from low yields. In order to ensure a stable and sustainable limonene supply, the development of a biotechnological process for limonene synthesis complements the traditional production route. Moreover, such a process could serve as a basis for the production of other monoterpenes of interest and subsequent selective functionalization.

During recent years, recombinant microbial strains have been engineered for limonene synthesis [11]. The production of isoprenoids with bacterial hosts was challenged by the low supply of the common precursors isopentenyl pyrophosphate (IPP) and dimethylallyl pyrophosphate (DMAPP) via the native 2-C-methyl-D-erythritol 4-phosphate (MEP) pathway. Higher precursor availability was realized by the introduction of a heterologous mevalonate (MVA) pathway from *Saccharomyces cerevisiae* in *Escherichia coli*, and isoprenoid titers above 100 mg·L^{-1} were achieved for the first time. A nine-enzyme pathway was constructed on three plasmids to produce amorpha-4,11-diene, which is the sesquiterpene precursor to artemisinin, an antimalarial drug [12]. Based on this study, an equivalent set of plasmids was designed to produce limonene with recombinant *E. coli* [8]. The pathway was optimized by balancing the involved enzymes in several iterative steps, and the number of plasmids was reduced to a single plasmid (Figure 1). Cultivations of the engineered *E. coli* strain in shake flasks using glucose as carbon source in a complex medium resulted in limonene titers of up to 400 mg·L^{-1}.

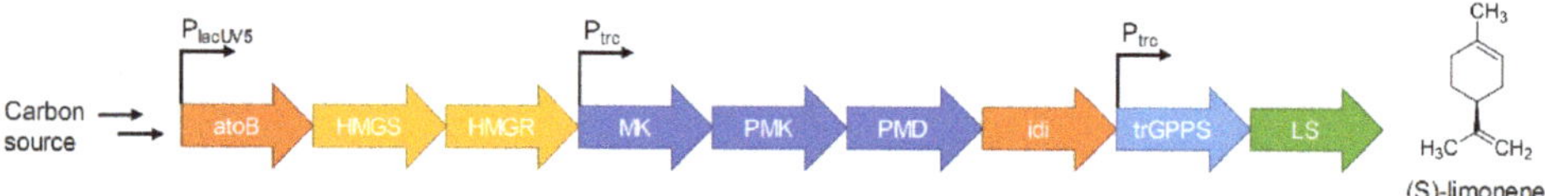

Figure 1. The heterologous mevalonate (MVA) pathway and limonene synthase introduced into *Escherichia coli* for the production of (S)-limonene. Acetoacetyl-CoA synthase from *E. coli* (atoB), HMG-CoA (hydroxymethylglutaryl-CoA) synthase from *Saccharomyces cerevisiae* (HMGS), an N-terminal truncated version of HMG-CoA reductase from *S. cerevisiae* (HMGR), mevalonate kinase (MK), phosphomevalonate kinase (PMK), phosphomevalonate decarboxylase from *S. cerevisiae* (PMD), isopentenyl diphosphate isomerase from *E. coli* (idi), a truncated and codon-optimized version of geranyl pyrophosphate synthase from *Abies grandis* (trGPPS), and a truncated and codon-optimized version of limonene synthase from *Mentha spicata* without the plastidial targeting sequence (LS).

Willrodt et al. constructed another *E. coli* strain harboring a two-plasmid system (pBAD:LS, pET24:AGPPS2) and operated a two-liquid phase fed-batch setup with a minimal medium in a stirred-tank bioreactor [13]. In this study, the addition of an inert organic phase was used to prevent product inhibition, toxicity effects, and the evaporative loss of limonene. Diisononyl phthalate (DINP) was selected as a biocompatible organic carrier solvent because of its favorable partition coefficient and lack of detectable impact on the growth of *E. coli* [14]. Final limonene concentrations of 1350 mg·L^{-1} were reached with glycerol as the sole carbon source, which was an almost 4-fold increase in limonene formation compared to that from glucose fermentations using the same strain. The use of glycerol resulted in a prolonged growth and production phase, leading to a more stable process with a maximum space-time yield of about 40 mg·L^{-1}·h^{-1} for carbon-limited cultivation [13].

Rational strain optimization, as well as reaction engineering, demonstrated the potential of the biotechnological production of monoterpenes. Nevertheless, space-time yields and product titers are still not applicable for industrial production. Additionally, data obtained at the bioreactor scale are rare. This study aims at the development of a feasible bioreactor scale process for monoterpene production with a recombinant *E. coli* strain that is genetically optimized for limonene synthesis.

2. Results

2.1. Influence of Inducer Concentration on Limonene Yields

Previous studies with a single plasmid strain (*E. coli* DH1 pJBEI-6409) cultivated in complex medium elucidated that low inducer concentrations (0.025 mM isopropyl β-D−1-thiogalactopyranoside (IPTG)) resulted in the highest limonene titers [8]. It was hypothesized that the amount of LacI produced by the single copy of *lacI* in the vector might not be enough to fully repress all three promoters in pJBEI-6409. The fully expressed MVA and limonene pathway at high IPTG levels could be too stressful for efficient limonene production. In the present study, different IPTG levels (0.025, 0.05, 0.1, 0.2, 0.5, and 1 mM) were tested for the optimal expression of heterologous genes. In comparison to the mentioned study, a different single plasmid strain was used (*E. coli* BL21 (DE3) pJBEI-6410), which carries a version of pJBEI-6409 harboring ampicillin resistance instead of chloramphenicol resistance. Furthermore, fermentations were carried out in M9 minimal medium instead of a complex medium. It turned out that the highest biomass specific yields could be obtained with IPTG concentrations of 0.05 mM and 0.1 mM (Figure 2). These values are high compared to the reported inducer concentrations for the producer strain *E. coli* DH1 pJBEI-6409 [8]. Following the hypothesis of Alonso-Gutierrez et al., the higher optimal inducer levels could be explained by a higher *lacI* expression level [8]. In contrast to *E. coli* DH1, the host strain *E. coli* BL21 (DE3) carries a Lac regulatory construct in its genome [15]. This operon includes *lacI*^{q}, which is a mutant of *lacI* with a 10-fold higher expression level that leads to a lower basal expression of T7 RNA and therefore to a more tightly controlled expression [16]. For the following experiments, the inducer concentration of 0.1 mM IPTG was chosen to ensure sufficient induction during bioreactor experiments.

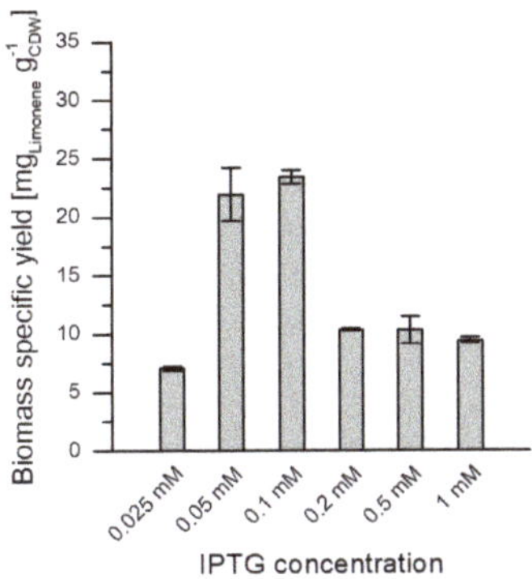

Figure 2. Biomass specific yields for different concentrations of the inducer isopropyl β-D−1-thiogalactopyranoside (IPTG) after 12 h of cultivation. Two-liquid phase shake flask fermentations with *E. coli* BL21 (DE3) pJBEI-6410 in M9 minimal medium with 0.5% *w/v* glucose as the sole carbon source. The error bars relate to biological duplicates.

2.2. Glycerol as the Sole Carbon Source for Fermentative Limonene Production

A prolonged growth phase of *E. coli* and higher product concentrations with glycerol as the sole carbon source were described by Willrodt et al. in an aforementioned study [13]. In order to investigate if these influences can also be observed with *E. coli* BL21 (DE3) pJBEI-6410, shake flask experiments were carried out using either glucose or glycerol as the sole carbon source. In the case of glucose, the substrate was consumed completely after 10 h, whereas glycerol was still present in the fermentation medium after 11 h. Finally, the carbon source was fully consumed in both cultivations. The growth curves were also similar (Appendix, Figure A1). Using glucose for the carbon supply resulted in a final limonene concentration of 121 ± 1 mg·L_{org}^{-1} in the organic phase (Figure 3A). By comparison, the fermentation with glycerol showed a prolonged production phase, resulting in a final limonene concentration of 184 ± 11 mg·L_{org}^{-1}. The limonene yields relative to the carbon source were 9.3 ± 0.1 g·C-mol^{-1} and 14.2 ± 0.8 g·C-mol^{-1}, for glucose and glycerol, respectively (Figure 3B). These results confirm previous observations that glycerol is the better choice as a carbon source for fermentative limonene production with *E. coli*.

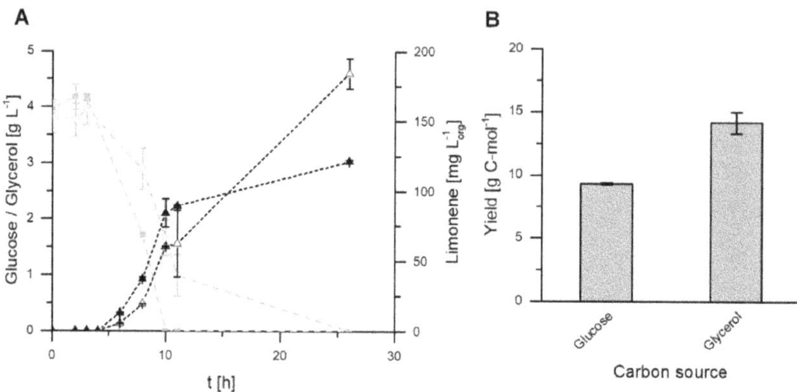

Figure 3. Two-liquid phase shake flask fermentations with *E. coli* BL21 (DE3) pJBEI-6410 in M9 minimal medium with either 0.5% *w/v* glucose (closed symbols) or glycerol (open symbols) as the sole carbon source. (**A**) Limonene concentrations (▲, △) in the organic phase and carbon source (■, □) concentrations were determined at regular intervals. (**B**) Carbon specific limonene yields after 26 h of cultivation. The error bars relate to biological duplicates.

2.3. Fermentative Limonene Production in a Stirred-tank Reactor

The first attempt to produce limonene with *E. coli* BL21 (DE3) pJBEI-6410 and glycerol as the sole carbon source in a two-liquid phase fed-batch setup was carried out in a 3.1 L reactor with 1 L of M9 minimal medium and 0.5 L of the organic carrier solvent diisononyl phthalate (DINP). A carbon limited exponential feed was applied with a calculated growth rate of 0.18 h^{-1} (Figure 4A,B), which was based on a substrate specific biomass production rate obtained from an initial batch cultivation. Heterologous gene expression was induced with the addition of 0.1 mM IPTG after two hours of fed-batch cultivation. After 10 h of exponential growth, no further increase in biomass was observed. The growth rate for this period was 0.15 h^{-1} and a final cell dry weight (CDW) of 28.7 g·L^{-1} was achieved. No accumulation of glycerol was detected until this time point. Acetate formation was suppressed during growth. After 11 h, growth stopped, and the exponential feed was set constant. During the next 15 h, glycerol accumulated, followed by acetate formation, leading to final concentrations of 5.7 g·L^{-1} glycerol and 2.8 g·L^{-1} acetate. The ammonium concentration increased during the fermentation from 0.4 to 1.9 g·L^{-1}, probably due to the pH regulation with ammonia as a result of acetate and carbon dioxide formation. The specific activity of limonene synthesis increased after induction with IPTG and reached a maximum of 2.6 U g_{CDW}^{-1} by the end of exponential growth. However, significant limonene formation was still observed for the non-growing cells, and a final limonene concentration of 4.4 g·L_{org}^{-1} was achieved. Concentrations of limonene were determined for the organic phase volume, because of the significant dilution of the aqueous phase due to the addition of the feed solution. The cell growth and the production of limonene seemed to be limited by a yet unidentified mechanism, which might be the accumulation of limonene or another compound related to the biosynthesis of limonene—such as an intermediate or metabolite—or a limitation caused by the depletion of a medium component.

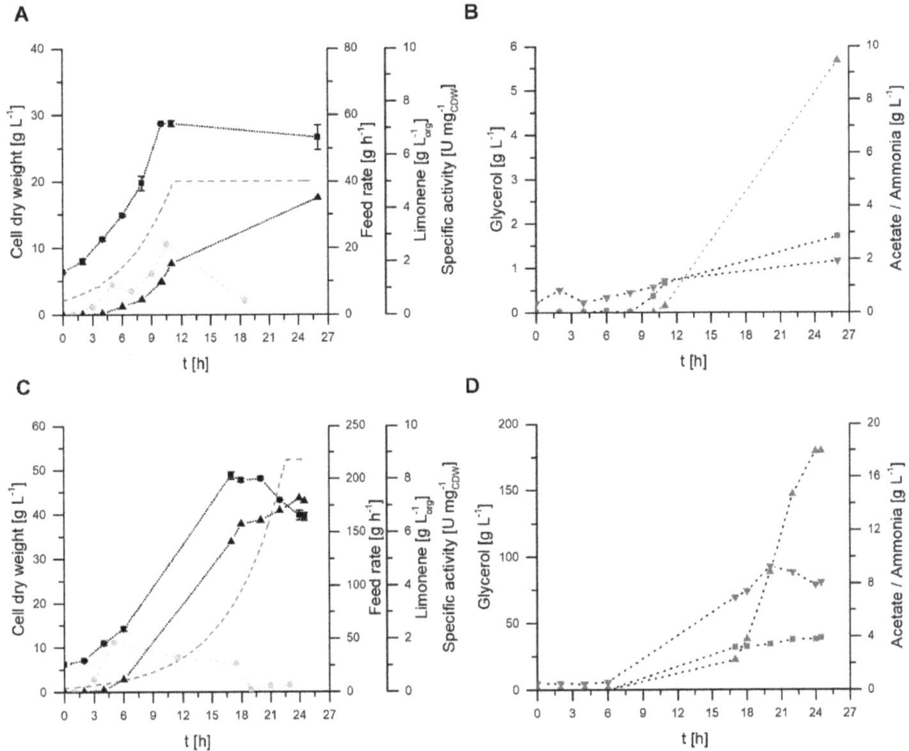

Figure 4. Two-liquid phase fed-batch fermentation with *E. coli* BL21 (DE3) pJBEI-6410 in M9 minimal medium. Cell dry weight (CDW) (■), limonene concentrations (▲) in the organic phase, glycerol (▲), acetate (▼), and ammonium (■) concentrations were determined at regular intervals. The specific activities (●) were calculated for distinct time points throughout the fermentation time. The feed rate is displayed as well (dotted line). (**A**) and (**B**) display the initial fed-batch fermentation (D = 0.18 h^{-1}), whereas (**C**) and (**D**) display the optimized fed-batch fermentation with a lower feed rate (D = 0.15 h^{-1}) and additional trace element supply. The error bars for CDW relate to two independent measurements.

To test whether intracellular limonene, terpene intermediates, or the expression of heterologous pathway genes influence cell growth, a glycerol-limited fed-batch fermentation was carried out without the induction of the heterologous pathway (data not shown). Although heterologous gene expression was not induced, 222 mg·L_{org}^{-1} limonene was produced during the exponential growth. These small amounts are probably caused by leaky expression of the pathway genes. Nearly identical CDW was achieved, while 20-fold less limonene was produced. Therefore, toxification by intermediates of the MVA pathway or intracellular limonene can be excluded.

In order to overcome the growth limitation, the parameters for the fermentation were changed to permit prolonged exponential growth, coupled with higher limonene concentrations. A carbon limited exponential feed was set up with a calculated growth rate of 0.15 h^{-1}, which is lower than in the previous experiments (0.18 h^{-1}) (Figure 4C,D). Furthermore, additional US* trace elements were supplied. After 6 and 16 h of cultivation, 2 and 1 mL of US* trace element solution were spiked, respectively. The exponential growth could be maintained for nearly 17 h, with a growth rate of 0.12 h^{-1}. Thus, the addition of trace elements prolonged the growth phase, and a CDW of 48.8 g·L^{-1} was reached, whereas the second trace element spike did not affect the cells in their stationary phase. The specific activity quickly increased reaching a maximum of 1.8 U g_{CDW}^{-1} after 5 h, before it decreased with ongoing fermentation. In contrast to in previous experiments, the specific activity could be maintained above 1 U g_{CDW}^{-1} for a time period of more than 17 h. The prolonged production phase and higher biomass concentration resulted in 7.3 g·L_{org}^{-1} limonene after 24 h by the end of the fermentation. To our knowledge, this is the highest limonene titer reported so far.

3. Discussion

3.1. Glycerol is a Suitable Carbon Source for Heterologous Limonene Production in Escherichia coli

The beneficial effect of glycerol as a sole or supplementary carbon source has been reported before for the fermentative production of carotenoids in MEP engineered *E. coli* strains [17–19], for sesquiterpenes [12] and for limonene [13] in MVA-engineered *E. coli* strains. The use of glycerol as a carbon source resulted in higher limonene formation rates, a prolonged growth phase, and increased stability compared to the same whole-cell biocatalyst growing on glucose [13].

In the present study, we were able to transfer this knowledge to a bioprocess with an optimized one plasmid strain and showed that, compared to glucose, glycerol is definitely the preferred carbon source for the production of limonene with *E. coli*. The transferability to the bioreactor scale was validated, and reaction engineering was performed to further increase the limonene titer. This led to the highest monoterpene concentration obtained with a microorganism to date. Next to availability and low cost, various advantages make glycerol an attractive carbon source for fermentation processes compared to glucose. Firstly, beneficial effects on the viability of cells and productivity of recombinant proteins were observed [20]. Secondly, glycerol does not show any catabolic repression in combination with lactose, which might be the preferred inducer for heterologous gene expression instead of IPTG due to a reduced stress level for the production host. Catabolite repression occurs when excess glucose is present and leads to reduced lactose uptake rates, which causes the decreased expression of recombinant proteins [21]. Finally, glycerol is a suitable carbon source for anaerobic fermentation with *E. coli* strains producing biofuels and highly reduced compounds. The high degree of reduction of carbon atoms in glycerol ($\kappa = 4.67$) provides a distinct advantage over glucose ($\kappa = 4.00$) in the absence of other electron acceptors [22]. *E. coli* strains are able to utilize glycerol in such conditions for cell growth and need a suitable sink for the excess reducing equivalents generated during the formation of biomass [23]. Therefore, the ability to form a highly reduced product is essential for the microorganism. Limonene has a high degree of reduction ($\kappa = 5.60$), so it would be a suitable product and sink for reducing equivalents in anaerobic glycerol fermentation. The anaerobic environment could have another beneficial effect regarding the toxicity of limonene. Whereas limonene itself has relatively little toxicity towards *E. coli* cells, the common oxidation product limonene hydroperoxide, which forms spontaneously in aerobic environments, shows highly antimicrobial effects [24]. In this study, the inhibitory effects of limonene hydroperoxide were not observed, due to efficient product extraction in the organic phase.

3.2. Progess to an Economic Limonene Production Process

The optimized bioreactor process described in this study resulted in limonene productivity exceeding the threshold for developing a profitable production process for fine chemicals (100 mg·L^{-1}·h^{-1}) [25] for the first time (Table 1). However, a techno-economic assessment stated that a biotechnological production process for limonene needs to have a space-time yield above 700 mg·L^{-1}·h^{-1} and a 45% carbon specific yield to be competitive with established processes [26]. While our process already shows more than a fifth of the space-time yield required, the conversion of the carbon source into the product is still low, at less than 1%.

Table 1. *Escherichia coli*-based fermentation setups for the production of limonene.

Strain	Plasmid System	Medium	Setup	$c_{Limonene}$ [mg·L^{-1}]	STY [mg·L^{-1} h^{-1}]	Reference
E. coli DH1	pJBEI-6409	EZ-Rich / glucose	Batch—Shake flask	435	6	[8]
E. coli BL21 (DE3)	pBAD:LS, pET24:AGPPS2	M9 / glycerol	Fed-batch—STR	1350	23	[13]
E. coli BW25113 (DE3)	pMAP6, pISP6, pNLSt1	YM9 / glucose	Fed-batch—Shake flask	1290	15	[27]
E. coli BL21 (DE3)	pJBEI-6410	M9 / glycerol	Fed-batch—STR	3630	151	This study

STY: space-time yield; STR: stirred tank reactor.

Different approaches to improve the yield have already been described. Non-growing but metabolically active *E. coli* cells can boost the production of the desired product due to reduced energy and carbon loss to biomass formation [28]. A fourfold increase in specific limonene yields relative to biomass was accomplished with this strategy. Moreover, further pathway debottlenecking and the optimization of the involved enzymes could increase the economics of the process. For example, the exchange of the geranyl pyrophosphate synthase with a neryl pyrophosphate synthase from *Solanum lycopersicum* led to increased limonene production with *E. coli* [27]. Another strategy to increase the product titer and the specificity of monoterpenes in *E. coli* was described by Chacón et al. [29]. The monoterpene geraniol was converted to the monoterpenoid geranyl acetate with an in vivo esterification and extracted in situ to an organic phase. Toxicity issues and the synthesis of by-products could be circumvented, resulting in a monoterpenoid concentration of 4.8 $g{\cdot}L^{-1}$. Similar approaches to the coupled synthesis and functionalization of limonene are described, which produce the valuable monoterpenoid perillyl alcohol [8,30].

The highest yields with more than 95% and titers of 15 $g{\cdot}L^{-1}$ were achieved with a cell-free system consisting of 27 purified enzymes, which convert glucose into monoterpenes [31]. Other systems which incorporate acetic acid as a starting building block for the cell-free synthesis of terpenes are described as well [32]. However, a major drawback is the need for purified enzymes, which are associated with additional costs and the further input of glucose needed to produce them. The need for enzyme purification can be avoided with the use of enzyme-enriched *E. coli* lysates, but this approach appeared to suffer from low product titers of 90 $mg{\cdot}L^{-1}$ [33]. Moreover, the involved enzymes are considered to have low stabilities in the in vitro environment, and cofactor regeneration could be a limiting aspect in cell-free applications as well [32]. The expensive cofactors CoA and NADPH must be effectively recycled in such systems, while the use of whole cells circumvents these drawbacks as the cofactors are regenerated by the primary metabolism. Therefore, microorganisms are preferred as the biocatalyst for the larger biotechnological production of limonene. Microbial hosts other than *E. coli* were recently investigated as producer strains, such as the cyanobacterium *Synechocystis* sp. [34] or the oleaginous yeast *Yarrowia lipolytica* [35], which was able to produce limonene from waste cooking oil. However, product titers were orders of magnitude lower compared to the processes based on engineered *E. coli*.

Next to the selection and optimization of the production system, a feasible bioprocess with high limonene titers involves the integrated development of in situ product removal strategies. Due to the high volatility and inhibitory effects on cell growth, the capturing of limonene during fermentation is required. Various methods are available, with two-liquid phase and gas stripping systems being especially suitable at higher scales [36]. In particular, two-liquid phase systems have the advantage that the products are effectively removed from the fermentation broth [37]. The choice of capturing method is also dependent on the further application of the product. If limonene is subsequently used as a pure compound, solvent-free systems might be the better choice, whereas application as an additive for, e.g., solvents might allow the use of the same solvent for in situ extraction [38]. In the present study, the in situ product removal strategy in combination with an engineered *E. coli* strain and a glycerol-limited fed-batch fermentation enabled the synthesis of the highest limonene concentration reported to date. Steps towards an economic process were made, and the potential of integrating already generated knowledge with the biotechnological production of terpenes was demonstrated.

4. Materials and Methods

4.1. Chemicals and Bacterial Strains

All chemicals used in this work were purchased from Carl Roth GmbH & Co. KG (Karlsruhe, Germany) and Merck KGaA (Darmstadt, Germany).

E. coli BL21 (DE3) harboring the plasmid pJBEI-6410 was used for all experiments. pJBEI-6410 carries the genes for the MVA pathway, a geranyl pyrophosphate synthase, and the limonene synthase from *Mentha spicata* [8]. It was a gift from Taek Soon Lee (RRID:Addgene_47049; http://n2t.net/addgene:47049).

4.2. Cultivation and Fermentative Limonene Production in Shake Flasks

For the cultivation in shake flasks, precultures were grown in 5 mL of LB medium (10 $g{\cdot}L^{-1}$ tryptone, 5 $g{\cdot}L^{-1}$ yeast extract, and 5 $g{\cdot}L^{-1}$ NaCl) with 100 $\mu g{\cdot}mL^{-1}$ ampicillin for 4 to 6 h at 37 °C and shaking at 200 rpm. LB precultures were transferred 1:500 to 250 mL baffled Erlenmeyer flasks with 50 mL M9 minimal medium (8.5 $g{\cdot}L^{-1}$ $Na_2HPO_4 \cdot 2H_2O$, 3 $g{\cdot}L^{-1}$ KH_2PO_4, 0.5 $g{\cdot}L^{-1}$ NaCl, 1 $g{\cdot}L^{-1}$ NH_4Cl, 2 mL of 1 M $MgSO_4$, and 1 mL of L^{-1} US* trace element solution) with either 0.5% *w*/*v* glucose or 0.5% *w*/*v* glycerol as the sole carbon source. The cultures were incubated overnight at 30 °C with shaking at 200 rpm. The M9 precultures were used to inoculate 80 mL of M9 minimal medium with the 0.5% *w*/*v* carbon source in 500 mL baffled shake flasks to an optical density of 0.11 at 600 nm (OD_{600}). The main cultivations were performed at 30 °C with shaking at 200 rpm. Heterologous gene expression was induced by adding 0.1 mM isopropyl β-D-1-thiogalactopyranoside (IPTG) once an OD_{600} of 0.4–0.6 was reached. After induction, the cultures were overlaid with 20 mL of diisononyl phthalate (DINP), and incubation was continued. For sampling, the phases were separated by centrifugation (2 min, 4 °C, 11,000 × *g*). Prior to gas chromatography (GC) analysis, the organic phase was diluted in diethyl ether with 0.2 mM dodecane and dried over anhydrous Na_2SO_4. The aqueous phase was used for HPLC analysis. The cell dry weight was determined by measurement of the optical density at a wavelength of 600 nm (Libra S11 Visible Spectrophotometer, Biochrom, Cambridge, UK). One OD_{600} unit corresponded to 0.312 g_{CDW} L^{-1}, with a linear range between 0.1 and 1.

4.3. Fermentative Limonene Production in a Stirred-Tank Reactor

Two-liquid phase fermentations at the bioreactor scale were carried out in a 3.1 L stirred-tank reactor (KLF 2000, Bioengineering AG, Wald, Switzerland). The tank reactor was equipped with two Rushton turbine stirrers. An M9 preculture was used to inoculate an initial batch cultivation, which was performed in 1 L of M9 minimal medium containing a 1.5% *w*/*v* carbon source at a starting OD_{600} of 0.11. The pH was set to 7.2 and controlled by the automatic addition of either 30% *v*/*v* phosphoric acid or 25% *v*/*v* ammonia hydroxide solution. Batch cultures were incubated at 30 °C, with shaking at 1800 rpm and an aeration rate of 1 vvm, until the batch phase was finished after 13 to 17 h (indicated by a steep pO_2 increase). Before the start of the carbon-limited fed-batch fermentation, 1 mL of US* trace element solution was added. To keep the exponential growth rates between 0.18 and 0.2 h^{-1}, a carbon-limited and controlled exponential feed with a 73% *w*/*v* carbon source and 19.6 $g{\cdot}L^{-1}$ $MgSO_4 \cdot 7\,H_2O$ was started. The pO_2 value was kept above 30% by adjusting the stirrer speed and the aeration rate. If desired, recombinant gene expression was induced by adding 0.1 mM IPTG after 2 h of exponential feeding, and 0.5 L of DINP were added subsequently. Antifoam A was added only in cases of excessive foaming. Regular sampling was carried out as described above.

4.4. Quantification of Limonene

The quantification of limonene in DINP was carried out with GC using a TRACE GC Ultra (Thermo Fisher Scientific Inc., Waltham, MA, USA), equipped with a FactorFour-5ms column (30 m, 0.25 mm, 0.25 µm, Varian, Inc., Palo Alto, CA, USA) and a flame ionization detector. Nitrogen was used as a carrier gas and the injection volume was set to 1 µL (80 °C, 5 min; 80–140 °C, 7.5 $°C{\cdot}min^{-1}$; 140–300 °C, 40 $°C{\cdot}min^{-1}$; 300 °C, 5 min). Samples of the organic phase were prepared in diethyl ether containing 0.2 mM dodecane as an internal standard. The quantification of limonene was performed using a standard curve of the ratio between commercial (S)-limonene (Merck KGaA, Darmstadt, Germany) and the internal standard.

4.5. Quantification of Glucose, Glycerol, Acetate, and Ammonia

The quantifications of glucose, glycerol, and acetate concentrations were performed by high performance liquid chromatography (HPLC) with a LaChrome Elite HPLC system (VWR, Darmstadt, Germany), equipped with a Trentec 308R-Gel.H column (Trentec Analysetechnik, Gerlingen, Germany) and a refractive index detector. The mobile phase consisted of 5 mM sulphuric acid. An isocratic method was used, with a flow rate of 1 mL min^{-1} and an injection volume of 20 μL. The column oven was set to 40 °C. Eluted components were quantified using standard curves for glucose, glycerol, and acetate.

Ammonia concentrations were determined by the method of Berthelot [39].

Author Contributions: Conceptualization, M.K.J. and S.L.; methodology, J.R. and M.K.J.; formal analysis, J.R.; investigation, J.R.; resources, S.L.; data curation, K.R., M.K.J. and S.L.; writing—original draft preparation, J.R. and K.R.; writing—review and editing, K.R., M.K.J. and S.L.; visualization, J.R.; supervision, M.K.J. and S.L.; project administration, M.K.J. and S.L. All authors have read and agreed to the published version of the manuscript.

Funding: This research received no external funding.

Acknowledgments: We acknowledge financial support by Deutsche Forschungsgemeinschaft and TU Dortmund University within the funding program Open Access Publishing. We gratefully thank Christian Willrodt for fruitful scientific discussions and Saikai Humaerhan for her support in initial growth studies.

Conflicts of Interest: The authors declare no conflict of interest. The funders had no role in the design of the study; in the collection, analyses, or interpretation of data; in the writing of the manuscript, or in the decision to publish the results.

Appendix A

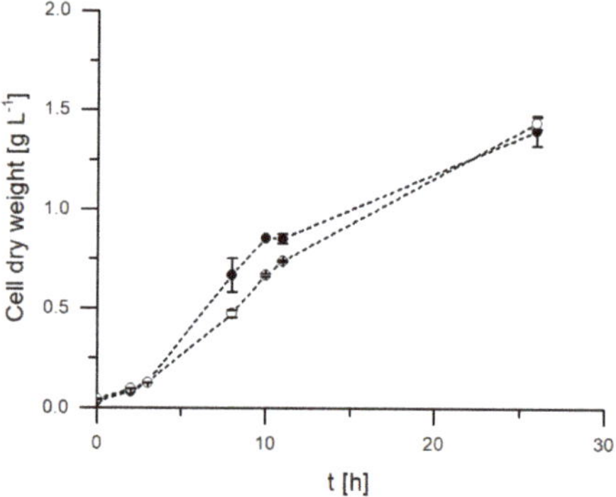

Figure A1. Two-liquid phase shake flask fermentations with *E. coli* BL21 (DE3) pJBEI-6410 in M9 minimal medium with either 0.5% *w/v* glucose (closed symbols, •) or glycerol (open symbols, ○) as the sole carbon source. Cell dry weights were determined at regular intervals. The error bars relate to biological duplicates.

References

1. Fisher, K. Terpenes replacing BTEX in oil field. Available online: http://www.aogr.com/magazine/editors-choice/terpenes-replacing-btex-in-oil-field (accessed on 12 February 2020).
2. Renninger, N.S.; Ryder, J.; Fisher, K. Jet fuel compositions and methods of making and using same. WO2008133658, 6 November 2008.
3. Chuck, C.J.; Donnelly, J. The compatibility of potential bioderived fuels with Jet A-1 aviation kerosene. *Appl. Energy* **2014**, *118*, 83–91. [CrossRef]
4. Zhang, J.; Zhao, C. A new approach for bio-jet fuel generation from palm oil and limonene in the absence of hydrogen. *Chem. Commun.* **2015**, *51*, 17249–17252. [CrossRef] [PubMed]

5. Inouye, S.; Takizawa, T.; Yamaguchi, H. Antibacterial activity of essential oils and their major constituents against respiratory tract pathogens by gaseous contact. *J. Antimicrob. Chemother.* **2001**, *47*, 565–573. [CrossRef] [PubMed]
6. Thomas, A.F.; Bessière, Y. Limonene. *Nat. Prod. Rep.* **1989**, *6*, 291–309. [CrossRef]
7. Cornelissen, S.; Julsing, M.K.; Volmer, J.; Riechert, O.; Schmid, A.; Bühler, B. Whole-cell-based CYP153A6-catalyzed (S)-limonene hydroxylation efficiency depends on host background and profits from monoterpene uptake via AlkL. *Biotechnol. Bioeng.* **2013**, *110*, 1282–1292. [CrossRef]
8. Alonso-Gutierrez, J.; Chan, R.; Batth, T.S.; Adams, P.D.; Keasling, J.D.; Petzold, C.J.; Lee, T.S. Metabolic engineering of *Escherichia coli* for limonene and perillyl alcohol production. *Metab. Eng.* **2013**, *19*, 33–41. [CrossRef]
9. Belanger, J.T. Perillyl Alcohol: Applications in Oncology. *Altern. Med. Rev.* **1998**, *3*, 448–457.
10. Rosenthal, K.; Lütz, S. Recent developments and challenges of biocatalytic processes in the pharmaceutical industry. *Curr. Opin. Green Sustain. Chem.* **2018**, *11*, 58–64. [CrossRef]
11. Kemper, K.; Hirte, M.; Reinbold, M.; Fuchs, M.; Brück, T. Opportunities and challenges for the sustainable production of structurally complex diterpenoids in recombinant microbial systems. *Beilstein J. Org. Chem.* **2017**, *13*, 845–854. [CrossRef]
12. Martin, V.J.J.; Pitera, D.J.; Withers, S.T.; Newman, J.D.; Keasling, J.D. Engineering a mevalonate pathway in *Escherichia coli* for production of terpenoids. *Nat. Biotechnol.* **2003**, *21*, 796–802. [CrossRef]
13. Willrodt, C.; David, C.; Cornelissen, S.; Bühler, B.; Julsing, M.K.; Schmid, A. Engineering the productivity of recombinant *Escherichia coli* for limonene formation from glycerol in minimal media. *Biotechnol. J.* **2014**, *9*, 1000–1012. [CrossRef] [PubMed]
14. Schewe, H.; Holtmann, D.; Schrader, J. P450BM-3-catalyzed whole-cell biotransformation of α-pinene with recombinant *Escherichia coli* in an aqueous–organic two-phase system. *Appl. Microbiol. Biotechnol.* **2009**, *83*, 849–857. [CrossRef] [PubMed]
15. Studier, F.W.; Moffatt, B.A. Use of bacteriophage T7 RNA polymerase to direct selective high-level expression of cloned genes. *J. Mol. Biol.* **1986**, *189*, 113–130. [CrossRef]
16. Calos, M.P. DNA sequence for a low-level promoter of the lac repressor gene and an 'up' promoter mutation. *Nature* **1978**, *274*, 762–765. [CrossRef]
17. Lee, P.C.; Mijts, B.N.; Schmidt-Dannert, C. Investigation of factors influencing production of the monocyclic carotenoid torulene in metabolically engineered *Escherichia coli*. *Appl. Microbiol. Biotechnol.* **2004**, *65*, 538–546. [CrossRef]
18. Kim, Y.S.; Lee, J.H.; Kim, N.H.; Yeom, S.J.; Kim, S.W.; Oh, D.K. Increase of lycopene production by supplementing auxiliary carbon sources in metabolically engineered *Escherichia coli*. *Appl. Microbiol. Biotechnol.* **2011**, *90*, 489–497. [CrossRef]
19. Jang, H.-J.; Yoon, S.-H.; Ryu, H.-K.; Kim, J.-H.; Wang, C.-L.; Kim, J.-Y.; Oh, D.-K.; Kim, S.-W. Retinoid production using metabolically engineered *Escherichia coli* with a two-phase culture system. *Microb. Cell Fact.* **2011**, *10*, 59. [CrossRef]
20. Kopp, J.; Slouka, C.; Ulonska, S.; Kager, J.; Fricke, J.; Spadiut, O.; Herwig, C. Impact of Glycerol as Carbon Source onto Specific Sugar and Inducer Uptake Rates and Inclusion Body Productivity in *E. coli* BL21(DE3). *Bioengineering* **2017**, *5*, 1. [CrossRef]
21. Wurm, D.J.; Hausjell, J.; Ulonska, S.; Herwig, C.; Spadiut, O. Mechanistic platform knowledge of concomitant sugar uptake in *Escherichia coli* BL21(DE3) strains. *Sci. Rep.* **2017**, *7*, 45072. [CrossRef]
22. Durnin, G.; Clomburg, J.; Yeates, Z.; Alvarez, P.J.J.; Zygourakis, K.; Campbell, P.; Gonzalez, R. Understanding and harnessing the microaerobic metabolism of glycerol in *Escherichia coli*. *Biotechnol. Bioeng.* **2009**, *103*, 148–161. [CrossRef]
23. Clomburg, J.M.; Gonzalez, R. Anaerobic fermentation of glycerol: A platform for renewable fuels and chemicals. *Trends Biotechnol.* **2013**, *31*, 20–28. [CrossRef] [PubMed]
24. Chubukov, V.; Mingardon, F.; Schackwitz, W.; Baidoo, E.E.K.; Alonso-Gutierrez, J.; Hu, Q.; Lee, T.S.; Keasling, J.D.; Mukhopadhyay, A. Acute limonene toxicity in *Escherichia coli* is caused by limonene hydroperoxide and alleviated by a point mutation in alkyl hydroperoxidase AhpC. *Appl. Environ. Microbiol.* **2015**, *81*, 4690–4696. [CrossRef] [PubMed]
25. Straathof, A.J.J.; Panke, S.; Schmid, A. The production of fine chemicals by biotransformations. *Curr. Opin. Biotechnol.* **2002**, *13*, 548–556. [CrossRef]

26. Sun, C.; Theodoropoulos, C.; Scrutton, N.S. Techno-economic assessment of microbial limonene production. *Bioresour. Technol.* **2020**, *300*, 122666. [CrossRef]
27. Wu, J.; Cheng, S.; Cao, J.; Qiao, J.; Zhao, G.-R. Systematic Optimization of Limonene Production in Engineered *Escherichia coli*. *J. Agric. Food Chem.* **2019**, *67*, 7087–7097. [CrossRef]
28. Willrodt, C.; Hoschek, A.; Bühler, B.; Schmid, A.; Julsing, M.K. Decoupling production from growth by magnesium sulfate limitation boosts de novo limonene production. *Biotechnol. Bioeng.* **2016**, *113*, 1305–1314. [CrossRef]
29. Chacón, M.G.; Marriott, A.; Kendrick, E.G.; Styles, M.Q.; Leak, D.J. Esterification of geraniol as a strategy for increasing product titre and specificity in engineered *Escherichia coli*. *Microb. Cell Fact.* **2019**, *18*, 105. [CrossRef]
30. Willrodt, C.; Hoschek, A.; Bühler, B.; Schmid, A.; Julsing, M.K. Coupling limonene formation and oxyfunctionalization by mixed-culture resting cell fermentation. *Biotechnol. Bioeng.* **2015**, *112*, 1738–1750. [CrossRef]
31. Korman, T.P.; Opgenorth, P.H.; Bowie, J.U. A synthetic biochemistry platform for cell free production of monoterpenes from glucose. *Nat. Commun.* **2017**, *8*, 15526. [CrossRef]
32. Dirkmann, M.; Nowack, J.; Schulz, F. An in Vitro Biosynthesis of Sesquiterpenes Starting from Acetic Acid. *ChemBioChem* **2018**, *19*, 2146–2151. [CrossRef]
33. Dudley, Q.M.; Nash, C.J.; Jewett, M.C. Cell-free biosynthesis of limonene using enzyme-enriched *Escherichia coli* lysates. *Synth. Biol.* **2019**, *4*, 1–9. [CrossRef] [PubMed]
34. Lin, P.-C.; Saha, R.; Zhang, F.; Pakrasi, H.B. Metabolic engineering of the pentose phosphate pathway for enhanced limonene production in the cyanobacterium *Synechocysti* s sp. PCC 6803. *Sci. Rep.* **2017**, *7*, 17503. [CrossRef] [PubMed]
35. Pang, Y.; Zhao, Y.; Li, S.; Zhao, Y.; Li, J.; Hu, Z.; Zhang, C.; Xiao, D.; Yu, A. Engineering the oleaginous yeast *Yarrowia lipolytica* to produce limonene from waste cooking oil. *Biotechnol. Biofuels* **2019**, *12*, 241. [CrossRef] [PubMed]
36. Jongedijk, E.; Cankar, K.; Buchhaupt, M.; Schrader, J.; Bouwmeester, H.; Beekwilder, J. Biotechnological production of limonene in microorganisms. *Appl. Microbiol. Biotechnol.* **2016**, *100*, 2927–2938. [CrossRef] [PubMed]
37. Xie, S.; Zhu, L.; Qiu, X.; Zhu, C.; Zhu, L. Advances in the Metabolic Engineering of *Escherichia coli* for the Manufacture of Monoterpenes. *Catalysts* **2019**, *9*, 433. [CrossRef]
38. Brennan, T.C.R.; Turner, C.D.; Krömer, J.O.; Nielsen, L.K. Alleviating monoterpene toxicity using a two-phase extractive fermentation for the bioproduction of jet fuel mixtures in *Saccharomyces cerevisiae*. *Biotechnol. Bioeng.* **2012**, *109*, 2513–2522. [CrossRef] [PubMed]
39. Searle, P.L. The berthelot or indophenol reaction and its use in the analytical chemistry of nitrogen. A review. *Analyst* **1984**, *109*, 549. [CrossRef]

Sample Availability: Samples of the compound (S)-limonene in DINP are available from the authors.

Article

A DyP-Type Peroxidase of *Pleurotus sapidus* with Alkene Cleaving Activity

Nina-Katharina Krahe *, Ralf G. Berger and Franziska Ersoy

Institut für Lebensmittelchemie, Gottfried Wilhelm Leibniz Universität Hannover, Callinstraße 5, 30167 Hannover, Germany; rg.berger@lci.uni-hannover.de (R.G.B.); franziska.ersoy@lci.uni-hannover.de (F.E.)
* Correspondence: nina.krahe@lci.uni-hannover.de; Tel.: +49-511-762-17257

Received: 20 February 2020; Accepted: 23 March 2020; Published: 27 March 2020

Abstract: Alkene cleavage is a possibility to generate aldehydes with olfactory properties for the fragrance and flavor industry. A dye-decolorizing peroxidase (DyP) of the basidiomycete *Pleurotus sapidus* (PsaPOX) cleaved the aryl alkene *trans*-anethole. The PsaPOX was semi-purified from the mycelium via FPLC, and the corresponding gene was identified. The amino acid sequence as well as the predicted tertiary structure showed typical characteristics of DyPs as well as a non-canonical Mn^{2+}-oxidation site on its surface. The gene was expressed in *Komagataella pfaffii* GS115 yielding activities up to 142 U/L using 2,2′-azino-bis(3-ethylbenzthiazoline-6-sulphonic acid) as substrate. PsaPOX exhibited optima at pH 3.5 and 40 °C and showed highest peroxidase activity in the presence of 100 μM H_2O_2 and 25 mM Mn^{2+}. PsaPOX lacked the typical activity of DyPs towards anthraquinone dyes, but oxidized Mn^{2+} to Mn^{3+}. In addition, bleaching of β-carotene and annatto was observed. Biotransformation experiments verified the alkene cleavage activity towards the aryl alkenes (*E*)-methyl isoeugenol, α-methylstyrene, and *trans*-anethole, which was increased almost twofold in the presence of Mn^{2+}. The resultant aldehydes are olfactants used in the fragrance and flavor industry. PsaPOX is the first described DyP with alkene cleavage activity towards aryl alkenes and showed potential as biocatalyst for flavor production.

Keywords: alkene cleavage; aryl alkenes; basidiomycota; biocatalysis; carotene degradation; dye-decolorizing peroxidase (DyP); manganese; *Komagataella pfaffii*; *Pleurotus sapidus*

---

1. Introduction

Many small aromatic aldehydes and ketones are volatiles with olfactory properties and therefore of high interest to the fragrance and flavor industry [1]. One method to generate aldehydes and ketones is the oxidative cleavage of alkenes. Chemical options are ozonolysis, dihydroxylation followed by oxidative glycol cleavage, or metal-based methods [2–4]. However, all of these methods have disadvantages, such as the generation of explosive intermediates, the use of environmentally unfriendly and/or toxic oxidants and metal catalysts, or low yield and low chemoselectivity [2]. An alternative is the application of enzymes due to their high chemo-, regio-, and stereospecificity as well as the possibility to use mild reaction conditions [3]. Another advantage is the generation of "natural" flavors according to effective legislation in Europe and the US. This becomes more and more important considering the rising popularity of natural products [5]. Different proteins of different enzyme classes, which are heme, non-heme iron, or non-iron metal dependent and have different protein structures as well as different reaction mechanism are known to catalyze alkene cleavage reactions [3]. Specifically, an isoeugnol and *trans*-anethole oxygenase from *Pseudomonas putida* and two manganese dependent enzymes from *Thermotoga maritima* (manganese-dependent Cupin TM1459) and *Trametes hirsuta* (Mn^{3+}-dependent proteinase A homologue) oxidatively cleaved the benzylic double bond of different aryl alkenes, such as isoeugenol and *trans*-anethole to form the respective aldehydes [6–9]. In addition, alkene cleavage activity towards aryl alkenes was also detected for several peroxidases. Cleavage of different styrene derivatives was described for *Coprinus cinereus* peroxidase and a human myeloperoxidase as minor side reaction [10], while horseradish peroxidase (HRP) showed a chemoselectivity of 92% for the conversion of *trans*-anethole (90%) to *p*-anisaldehyde [11]. Furthermore, transformations of *o*-ethylisoeugenol and *trans*-anethole to the corresponding benzaldehyde derivatives by lignin peroxidases were described [3,11]. However, to our best knowledge no alkene cleavage activity of a dye-decolorizing peroxidase (DyP) is known.

DyP-type peroxidases (EC: 1.11.1.19) are a new superfamily of heme peroxidases that oxidize various dyes, in particular xenobiotic anthraquinone dyes, which are hardly oxidized by other peroxidases [12]. Furthermore, typical peroxidase substrates, such as ABTS (2,2′-azino-bis(3-ethylbenzthiazoline-6-sulphonic acid) and phenolic compounds are also substrates for DyPs [13,14]. However, amino acid sequences, protein structures, and catalytic residues differ highly between DyPs and other classes of heme peroxidases [15]. Typical structural characteristics of DyPs are the ferredoxin-like fold, which is formed by two domains containing α-helices and four-stranded antiparallel β-sheets, and a GXXGD motif [14,15]. The active site (heme pocket) including a catalytic aspartic acid and arginine over the heme plane (distal) and proximal histidine is structurally similar to other heme peroxidases, even though other peroxidases contain histidine instead of aspartic acid [16]. The proximal histidine in the heme pocket functions as the fifth ligand of the heme iron, while the distal aspartic acid and arginine are involved in the activation of the enzyme [14,15]. The deprotonated aspartic acid (or asparagine) mediates the rearrangement of a proton from hydrogen peroxide after it enters the heme pocket in the resting state. This results in the heterolytic cleavage of hydrogen peroxide to water and oxidation of the heme to the radicalic-cationic oxoferryl species Compound I by two-fold single electron transfer [16]. Even though the distal arginine is not directly involved in the rearrangement of hydrogen peroxide, it is essential for the coordination of hydrogen peroxide to the heme iron and the stabilization of Compound I [16]. During the following reaction cycle, Compound I is reduced by oxidation of two substrate molecules to the state during enzyme resting state in two sequential steps with Compound II as intermediate. However, the existence of Compound II has not been confirmed universally for all DyP-type peroxidases [17]. In the presence of excessive hydrogen peroxide suicide inhibition was observed for different DyPs [18,19]. This is also well known for classical peroxidases as a result of an inactive oxidative state (Compound III) and results from reaction of hydrogen peroxide with Compound II [20,21].

The objective of the present study was to identify new enzymes of basidiomycetes with alkene cleavage activity towards aryl alkenes. A screening was performed using *trans*-anethole as model alkene. A new DyP-type peroxidase from *P. sapidus* was semi-purified and the coding gene was identified. Heterologous expression resulted in the production of soluble protein and allowed the biochemical characterization of the DyP. The enzyme was able to oxidize Mn^{2+}, but did not catalyze the degradation of anthraquinone dyes, which is typical for other DyPs. Biotransformation experiments verified the cleavage activity towards different alkenes. This is the first study describing a DyP with alkene cleavage activity towards aryl alkenes.

2. Results and Discussion

2.1. Purification and Identification of the Alkene Cleavage Activity

Within a screening of 17 basidiomycetes for alkene cleavage activity using the substrate *trans*-anethole (Table S1) *P. sapidus* turned out to be a promising candidate for the production of the desired activity. The lyophilized mycelium as well as the culture supernatant was examined for the ability to cleave *trans*-anethole after submerged cultivation. The culture supernatant showed no activity, whereas the incubation in the presence of the mycelium resulted in formation of 5.36 mM *p*-anisaldehyde (molar yield of 79.03%; Figure 1).

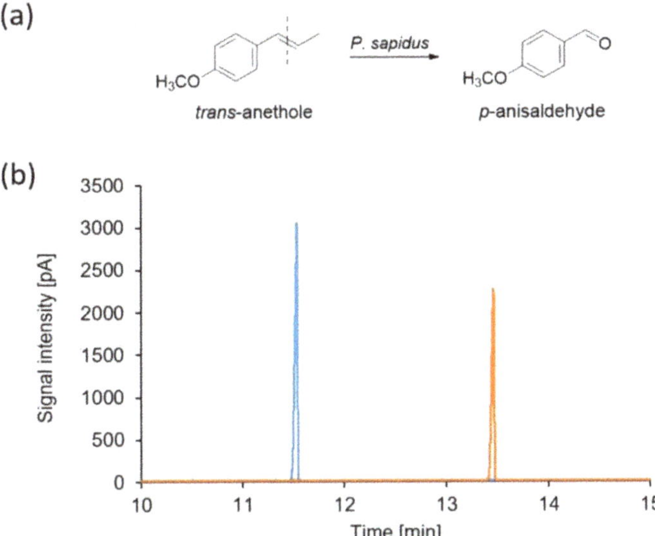

Figure 1. Bioconversion of *trans*-anethole by *P. sapidus*. (**a**) Alkene cleavage of *trans*-anethole resulted in the formation of *p*-anisaldehyde. (**b**) GC-FID chromatogram of an *n*-hexane extract of the conversion of the blank sample (blue) and after incubation with lyophilized mycelium of *P. sapidus* (orange). Retention times: *trans*-anethole (11.53 min) and *p*-anisaldehyde (13.45 min).

For the identification of the enzyme catalyzing the *trans*-anethole cleavage, it was semi-purified from the rehydrated mycelium by hydrophobic interaction and anion exchange chromatography (IEX). During the purification, a high activity loss occurred, which resulted in low product concentrations after conversion (Table S2). This was most likely a result of protein loss, enzyme degradation or denaturation as described for other enzymes [22–24]. Another possibility is the loss of cofactors or –substrates, such as metal ions or peroxides during the purification steps [6]. Addition of Mn^{2+} led to a 15-fold increase in *p*-anisaldehyde concentration, which was further increased by addition of hydrogen peroxide, indicating a cosubstrate dependency (Table S2). Chemical conversion by Mn^{2+} alone was excluded, while product formation was observed with hydrogen peroxide (18 µM), but with a yield around 30-fold lower than the one for the enzymatic reaction (Table S2). Thus, the improved bioconversion in the presence of Mn^{2+} and hydrogen peroxide was verified to be the result of an increased enzyme activity.

A class of fungal enzymes that requires hydrogen peroxide and some of which need Mn^{2+} for catalysis are peroxidases [12,25,26]. The anion exchange fractions, which showed alkene cleavage activity, also exhibited peroxidase activity, thus verifying the presence of a peroxidase. For visualization of the activity, a semi-native PAGE was performed and stained with ABTS in the presence of hydrogen peroxide (Figure 2). Two peroxidases running at 45 and 52 kDa were detected. The respective protein bands stained with Coomassie Brilliant Blue were excised for electrospray ionization tandem mass spectrometry. Due to the low protein concentration of the 52 kDa band, no meaningful peptides were found (Figure 2, lane 1). This paper presents the data obtained for the protein running at 45 kDa (Figure 2, arrow).

Three tryptic peptides (EGSELLGAR, DGSFLTFR, and SGAPIEITPLKDDPK) were identified by ESI-MS/MS. Homology searches against the public database NCBI using the mascot search engine (Matrix Science, London, UK) identified a DyP-type peroxidase of *P. ostreatus* PC15 (*Pleos*-DyP4 [27]; GenBank accession no. KDQ22873.1), a close relative of *P. sapidus*, as the best hit.

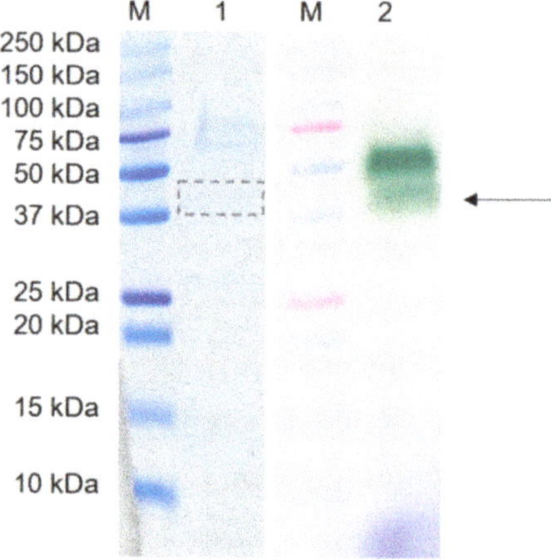

Figure 2. Semi-native PAGE of the active fraction after purification of the alkene cleavage enzyme from *P. sapidus* by IEX. 1: gel stained with Coomassie Brilliant Blue; 2: gel stained with ABTS in the presence of hydrogen peroxide, M: pre-stained molecular mass marker. An arrow and a box marks the protein that was successfully identified by sequencing.

2.2. Amplification and Expression of PsaPOX

Specific primers successfully amplified the 1512 bp coding region of the gene from *P. sapidus*. The translated amino acid sequence of 504 aa contained the peptide fragments that were obtained by ESI-MS/MS and showed highest identity (94%) to the sequence of *Pleos*-DyP4 (Figure 3), which has not been investigated for its alkene cleavage activity against *trans*-anethole or other substrates before [27,28]. In our hands, *P. ostreatus* showed a weaker *trans*-anethole cleavage activity, too (Table S1). Identity of the *P. sapidus* peroxidase (PsaPOX) to other DyPs and proteins was lower than 60%. A sequence alignment with other DyPs (Figure 3) confirmed that PsaPOX exhibited the typical GXXGD motif and all important residues known for the catalytic activity of DyPs, such as the proximal histidine (His-334) (fifth ligand of heme iron) and distal Asp-196 and Arg-360 involved in the activation of the enzyme (formation of compound I) by H_2O_2 cleavage [14,15]. Furthermore, Trp-405 was identified as a homolog to the surface exposed Trp-377 of AauDyP of *Auricularia auricula-judae*, which serves as an oxidation site for bulky substrates such as Reactive blue 19 (RBB19) using a long-range electron transfer [29]. Comparison of the PsaPOX and *Pleos*-DyP4 sequence indicated that PsaPOX exhibited a non-canonical Mn^{2+}-oxidation site on its surface (Asp-215, Glu-345, Asp-352 and Asp-354; Trp-339 participates in the electron transport from the oxidation site to the heme) like *Pleos*-DyP4 [28] and can oxidize Mn^{2+} to Mn^{3+}, which is known for a few fungal DyPs only [24,27,30,31].

A structural homology model of PsaPOX (Figure S1), which was generated using the X-ray crystal structure of *Pleos*-DyP4 (PDB-ID 6fsk) on the SWISS-MODEL server, possessed typical characteristics of the DyP-type peroxidase family (N- and C-terminal ferredoxin-like domain, each formed by four-stranded antiparallel β-sheets and several α-helices) [15] and supported the classification of PsaPOX as DyP. Furthermore, the analysis of the amino acid sequence using PeroxiBase [32] related the *P. sapidus* peroxidase (PsaPOX) to the class "DyP-type peroxidase D". As known from literature, DyPs differ significantly in amino acid sequence, tertiary structure, and catalytic residues from other representatives of the heme peroxidases, such as HRP, human myeloperoxidase, lignin peroxidase, or *Coprinus cinereus* peroxidase [12,15,33], all of which are able to cleave *trans*-anethole or structurally related alkenes [10,11]. So far no DyP is known to catalyze the mentioned reaction.

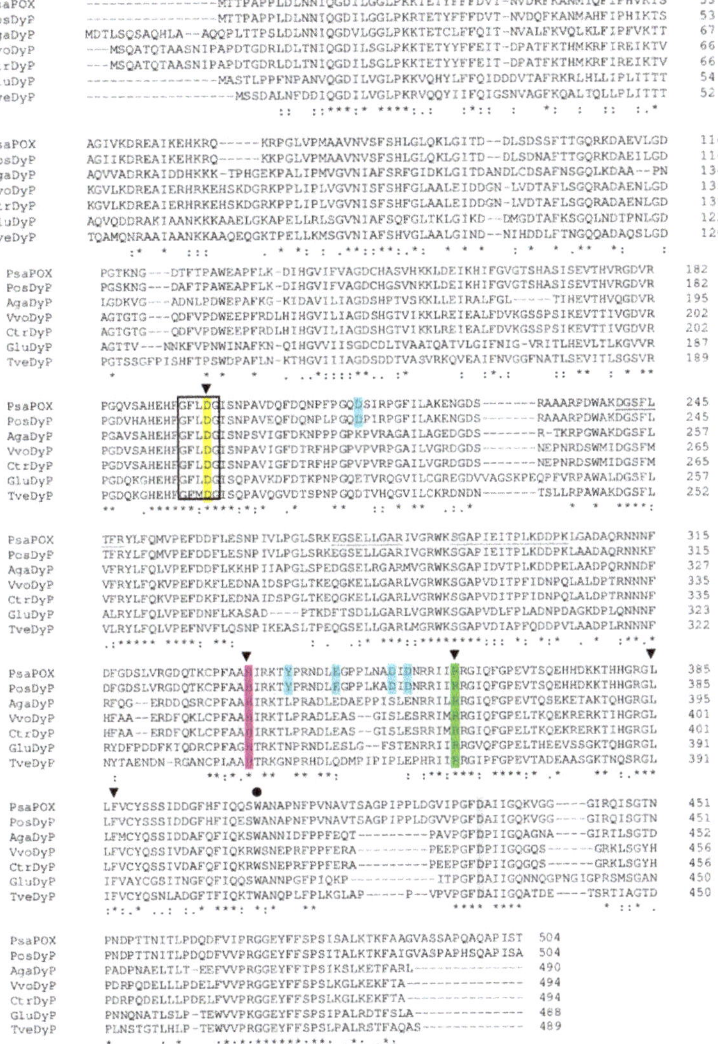

Figure 3. Alignment of alkene cleaving peroxidase from *P. sapidus* (PsaPOX) with the *Pleos*-DyP4 of *P. ostreatus* (PosDyP; KDQ22873.1) and other characterized DyPs. AgaDyP: *Armillaria gallica* (PBK80505.1), VvoDyP: *Volvariella volvacea* (AKU04643.1), CtrDyP: *Coriolopsis trogii* (AUW34346.1), GluDyP: *Ganoderma lucidum* (ADN05763.1), and TveDyP: *Trametes versicolor* (XP_008039377.1). Inverted triangles show amino acids important for heme binding (histidine (magenta) functions as ligand for heme and the four other amino acid residues form a hydrogen peroxide binding pocket). Aspartic acid, which forms a hydrogen bond with histidine to stabilize compound I (oxidized heme after transfer of two electrons to H_2O_2) is shown in grey. The black box indicates the GXXDG motif containing the catalytic aspartic acid residue (yellow), which cleaves H_2O_2 heterolytically with the help of the neighboring arginine (green) to form compound I, and the circle presents an exposed tryptophan potentially involved in an LRET (long range electron transfer). Important amino acids for Mn^{2+}-oxidation are highlighted in cyan; asterisks indicate conserved residues, colons equivalent residues and dots partial residue conservation. Peptides identified by protein sequencing are underlined. Alignment was performed with Clustal Omega (European Bioinformatics Institute, Hinxton, UK).

2.3. Production and Purification of the Recombinant PsaPOX

The *PsaPOX* gene was amplified and cloned into the *K. pfaffii* expression vector pPIC9. The initial expression of the gene yielded average peroxidase activities of 65 U/L after 72 h of cultivation. The best performing colonies produced activities up to 142 U/L, indicating a multiple insertion of the expression construct [34]. Similar results were obtained for the heterologous production of a DyP from *Funalia trogii* in *K. pfaffii* previously [24]. Further experiments were performed using the clone with the highest peroxidase activity for maximum protein production.

The recombinant peroxidase was purified by Ni-NTA affinity. Using SDS-PAGE, a molecular mass of around 61 kDa was determined (Figure 4a), which is slightly higher than the calculated molecular mass of 54.9 kDa (ExPASy). In addition, the native recombinant enzyme was detected at 52 kDa after semi-native PAGE, while the native wild-type enzyme showed a band at 45 kDa (Figures 2 and 4b). Deglycosylation by endoglycosidase H (EndoH) showed that the higher molecular mass was attributed to post-translational modifications by *K. pfaffii*, as has been described for other proteins [35,36]. The wild-type peroxidase, on the contrary, was not glycosylated (see Figures 2 and 4b). That is uncommon for DyPs, which usually exhibit a carbohydrate content of 9 to 30% [12]. The molecular mass of the monomeric PsaPOX was similar to other DyP-type peroxidases [12].

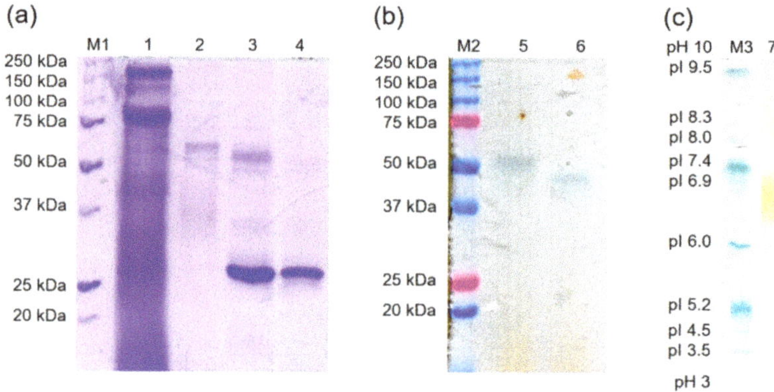

Figure 4. Purification of the recombinant PsaPOX by Ni-IMAC. (**a**) SDS-PAGE stained with Coomassie Brilliant Blue. 1: flow through, 2: elution fraction, 3: elution fraction incubated with EndoH, 4: EndoH, M1: molecular mass marker. (**b**) Semi-native PAGE stained with ABTS in the presence of hydrogen peroxide. 5: elution fraction, 6: elution fraction incubated with EndoH, M2: pre-stained molecular mass marker. (**c**) Isoelectric focusing gel. 7: elution fraction stained with phenylendiamine in the presence of urea peroxide. M3: standard protein marker for isoelectric focusing stained with Coomassie Brilliant Blue.

Analysis of the purified recombinant peroxidase by isoelectric focusing indicated an isoelectric point around pH 6.7 (Figure 4c), which differs slightly from the calculated value of 6.28 (ExPASy), but was similar to the isoelectric point of another DyP-type peroxidase from *P. sapidus* [19]. Most other proteins belonging to the DyP-type peroxidase family showed lower values (pI 3.5-4.3, [12]).

2.4. Biochemical Characterization of PsaPOX

The influence of pH and temperature on PsaPOX activity and stability was determined using ABTS in the presence of hydrogen peroxide as substrate (Figure 5). The enzyme showed a pH optimum of 3.5 while more than 50% of activity was conserved between pH 3 and 5 (Figure 5a). At lower or higher pH values of $\leq$ 25% of activity remained, most likely due to conformational changes of the enzyme. The results were consistent with the findings for other fungal DyPs, which had pH optima in the range between pH 2 and 5 [13,19,27]. PsaPOX showed the highest pH stability with a residual peroxidase activity of ≥90% between pH 2.0 and 5.5 after 1 h of incubation (Figure 5c). At pH values higher than six, near the isoelectric point, the stability decreased drastically, probably due to a reduced solubility and changes of the protein structure, which may have resulted in protein aggregation.

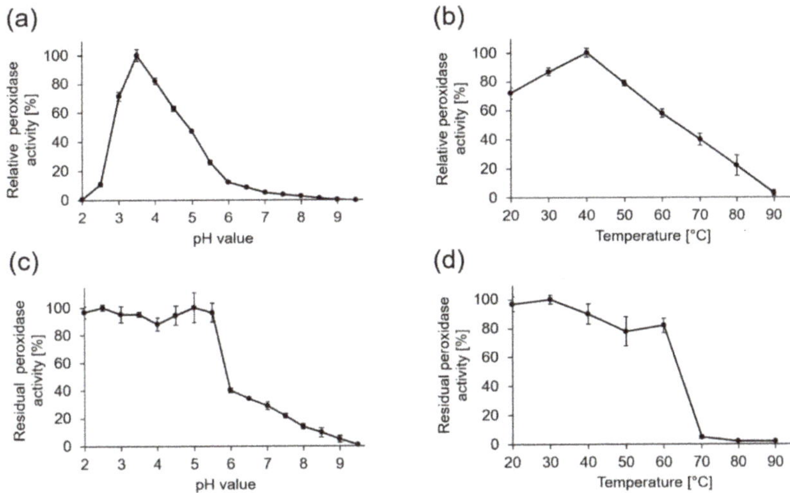

Figure 5. Influence of pH and temperature on activity and stability of PsaPOX. The pH optimum (**a**) was determined to be 3.5 and the temperature optimum (**b**) 40 °C. Relative peroxidase activity [%] was defined as the percentage of activity detected with respect to the highest activity in each experiment. pH stability (**c**) was determined after incubation of PsaPOX in Britton Robinson buffer ranging from pH 2.0 to 9.5 for 1 h at RT and temperature stability (**d**) after incubation at 20 to 90 °C and pH 3.5 for 1 h. Residual activities were determined at pH 3.5 and 40 °C. Values are the average of triplicate experiments with standard deviations shown as error bars.

Peroxidase activity of PsaPOX increased with rising temperature, reaching its maximum at 40 °C (Figure 5b), which was similar to the optimum (30–40 °C) of a recombinant DyP from *P. ostreatus* [13], but higher than the optimum (RT) of another DyP-type peroxidase of *P. sapidus* produced heterologously in *Escherichia coli* [37]. With further temperature increase, the peroxidase activity of PsaPOX decreased continuously. The temperature stability of PsaPOX was determined after an incubation for 1 h at different temperatures (Figure 5d). The enzyme was relatively stable at temperatures from 20 to 60 °C with a residual activity ≥80%. At higher temperatures, a high loss of activity was observed due to protein denaturation, resulting in residual activities <5%. The temperature stability of PsaPOX was higher than the stabilities of DyPs from *Bjerkandara adusta* and *Auricularia auricular-judae*, which were produced heterologously in *E. coli* (residual activity ≥80% and <5% after 1 h at 20–50 °C and 60 °C, respectively [13]), and of a DyP from *P. sapidus* produced in *Trichoderma reesei* (residual activity ≥80% and <65% after 5 min at 15-45 °C and 50 °C, respectively [19]).

As mentioned above, the addition of hydrogen peroxide as well as Mn^{2+} led to an increase of the product concentration for the biotransformation of *trans*-anethole using the lyophilized mycelium of *P. sapidus* containing the wild-type PsaPOX (Table S2). For this reason, the hydrogen peroxide and Mn^{2+} dependencies were examined for the recombinant enzyme using ABTS as substrate at optimal pH and temperature (Figure 6). As expected, no peroxidase activity was detectable without hydrogen peroxide. The activity rose with increasing peroxide concentration and reached its optimum in the presence of 100 μM H_2O_2 (Figure 6a). An increase of the hydrogen peroxide concentration led to a continuous activity decrease. Suicide inhibition in the presence of excess hydrogen peroxide is well known for classical peroxidases as a result of the formation of an inactive oxidative state (Compound III) by reaction of H_2O_2 and Compound II [20,21], even if the existence of Compound II has not been confirmed for DyP-type peroxidases universally [17]. However, inhibition of other DyPs in the presence of higher hydrogen peroxide concentrations has been reported [18,19].

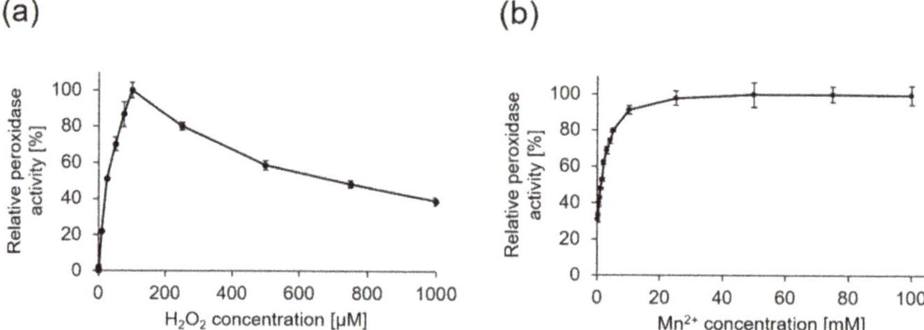

Figure 6. Effect of hydrogen peroxide (**a**) and Mn^{2+} concentration (**b**) on the activity of PsaPOX. Relative peroxidase activity [%] was defined as the percentage of activity detected with respect to the highest activity obtained in each experiment. Values are the average of triplicate experiments with standard deviations shown as error bars.

Investigation of the Mn^{2+} dependency (Figure 6b) showed that PsaPOX activity rose with increasing Mn^{2+} concentration, but was not completely dependent on the addition of Mn^{2+}. 30% of peroxidase activity were detected without addition of Mn^{2+}. PsaPOX reached the maximal activity in the presence of 25 mM Mn^{2+}. Evaluation of Mn^{3+} formation by Mn^{2+} oxidation revealed a manganese peroxidase activity of 0.4 U compared to 1 U of peroxidase activity using ABTS as substrate. This result fits the prediction of a Mn^{2+} oxidation site. Only a few fungal DyPs are known to catalyze the oxidation of Mn^{2+} [24,27,30,31]. Calculation of kinetic constants (Table 1) showed that the catalytic efficiency of PsaPOX towards Mn^{2+} was similar to the one of *Pleos*-DyP1 from *P. ostreatus* and Ftr-DyP from *Funalia trogii* [24,27]. However, the catalytic efficiency of *Pleos*-DyP4 was higher [27].

Table 1. Michaelis constants (K_m), catalytic constants (k_{cat}), and catalytic efficiencies (k_{cat}/K_m) for PsaPOX using ABTS, Mn^{2+}, Reactive blue 19 (RB19), and Reactive black 5 (RB5) as substrate. Values are the average of triplicate experiments with indication of standard deviations.

Substrate	K_m (µM)	k_{cat} (s^{-1})	k_{cat}/K_m (s^{-1} mM^{-1})
ABTS	37 ± 4	6.8 ± 0.2	184 ± 5
Mn^{2+}	1025 ± 79	7.2 ± 0.1	7 ± 0.1
RB19	n. d.	-	-
RB5	n. d.	-	-

n. d.: no activity was detected.

Kinetic parameters were also calculated for the oxidation of ABTS at optimal conditions (Table 1). The affinity of PsaPOX to ABTS (37 µM) was similar to a DyP from *Irpex lacteus* (28 µM), but higher in comparison to the FtrDyP from *F. trogii* (182 µM) and the *Pleos*-DyP2 from *P. ostreatus* (787 µM) [18,24,27]. In contrast, the catalytic efficiency of PsaPOX (184 s^{-1} mM^{-1}) was lower than the efficiency of the DyP from *Irpex lacteus* (8000 s^{-1} mM^{-1}) and *Pleos*-DyP4 (352 s^{-1} mM^{-1}), but higher than the efficiency of the FtrDyP (54 s^{-1} mM^{-1}).

It is known that DyP-type peroxidases typically oxidize anthraquinones and other dyes. Exemplary, decolorization of Reactive blue 19 (anthraquinone dye) and Reactive black 5 (recalcitrant azo dye) by recombinant PsaPOX (1 U/L) was tested. Unexpectedly, PsaPOX showed activity for neither of the substrates (Table 1), although the protein sequence and tertiary structure as well as the presence of typical catalytic residues and the GXXDG motif identified the enzyme as a DyP-type peroxidase. However, *Pleos*-DyP1 from *P. ostreatus* and *Tv*DyP1 from *Trametes versicolor* also did not oxidize the high redox-potential Reactive black 5, although they degraded Reactive blue 19 [27,30]. A missing activity against Reactive blue 19 or another anthraquinone has not been described for a fungal or class D type DyP before, but one bacterial DyP of *Pseudomonas fluorescens* (DyP2B, DyP typ class B) is known not to oxidize the anthraquinone dye Reactive blue 4 [38].

2.5. Alkene Cleavage Activity of PsaPOX

To prove the ability of PsaPOX to convert *trans*-anethole to *p*-anisaldehyde, biotransformation experiments were performed at optimal conditions using 1 U/mL peroxidase activity. Substrate cleavage was detected in the presence of hydrogen peroxide (Table S3), whereas no activity was observed in its absence as expected from the peroxidase activity measurement with ABTS (Figure 6a). Due to the fact that the semi-purified wild-type DyP showed an alkene cleavage activity without addition of hydrogen peroxide (see Table S2) a low amount of H_2O_2 must have been present in the analyzed IEX fraction. This was verified by incubation of the fraction with *o*-dianisidine and HRP in the presence of *trans*-anethole. Oxidation of *o*-dianisidine by HRP, which requires H_2O_2 as cosubstrate, and formation of a red-brown reaction product occurred (Figure S2). Further, analysis of the IEF fraction regarding a hydrogen peroxide producing enzyme revealed a hypothetical protein from *P. ostreatus* (KDQ29984.1). It belongs to the glucose-methanol-choline (GMC) oxidoreductase family as evident from the best hit for the protein band at 75 kDa (Figure 2, lane 1) according to protein sequencing (identified tryptic peptides: AADLIK, AIAVEFVR, ELGGVVDTELR, AQYDAWAELNR, VADASIIPIPVSAHTSSTVYMIGER, DLASGDPHGVGVSPESIDVTNYTR, VLGGSTTINAMLFPR, EVVVSAGTIGTPK) and homology search against the public database NCBI (Figure S3). The protein contained seven of the eight tryptic peptides identified for the 75 kDa band. The last one was found with an amino acid exchange (Arg instead of Lys), which is most likely a result of the different fungal strains the proteins originate from. The protein from *P. ostreatus* showed >92% identity to another hypothetical GMC oxidoreductase from *P. ostreatus* and ≥55% identity to a glucose oxidase from *Moniliophthora roreri* and other fungal alcohol oxidases (Figure S3), which belong to the GMC oxidoreductase family and are known for the production of hydrogen peroxide during substrate oxidation. Thus, the oxidase (75 kDa band) most likely produced the detected hydrogen peroxide, which was subsequently used as cosubstrate by the wild-type PsaPOX. Due to the fact that the formation of *p*-anisaldehyde by the oxidase under production of H_2O_2 seemed highly unlikely and as no further oxidation products of *trans*-anethole were detected, *trans*-anethole was excluded as substrate. Instead, the buffer component Bis-Tris, which contains several alcohol groups, or carbohydrate functionalities of other proteins in the IEX fraction were assumed to be used as substrate by the oxidase.

As described for the wild-type DyP the *p*-anisaldehyde concentration increased for the biotransformation with the recombinant enzyme in the presence of 25 mM Mn^{2+} (Table S3). However, product formation in general was low. The residual peroxidase activity was determined during the biotransformation of *trans*-anethole (Figure S4). After 16 h, 62% of the activity remained, thus inactivation of the enzyme was not responsible for the relatively low product yields.

PsaPOX (1 U/mL) was further examined for alkene cleavage activity regarding other substrates in the presence of hydrogen peroxide and Mn^{2+}. The aryl alkenes (*E*)-methyl isoeugenol as well as α-methylstyrene, which are derivatives of *trans*-anethole, were converted to the expected products (veratraldehyde and acetophenone), while piperine was not cleaved (Figure 7a). However, the resulting product concentration was fivefold lower for the biotransformation of (*E*)-methyl isoeugenol and more than tenfold lower for the conversion of α-methylstyrene than for *trans*-anethole. Different substrate specificities were also observed for the alkene cleavage by other peroxidases, such as HRP, *Coprinus cinereus* peroxidase, and a human myeloperoxidase [10,11], but a conversion of aryl alkenes using a DyP-type peroxidase has not been described before.

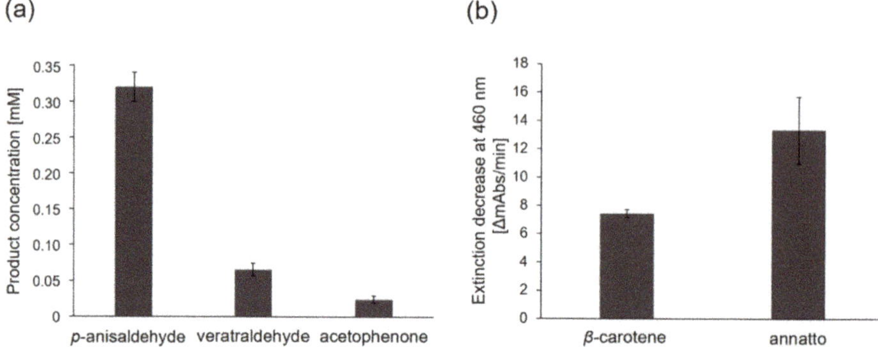

Figure 7. Alkene cleavage activity of PsaPOX on different substrates in the presence of 100 μM H_2O_2 and 25 mM $MnSO_4$ at pH 3.5. (**a**) Product concentration after conversion of *trans*-anethole (6.7 mM) to *p*-anisaldehyde, (*E*)-methyl isoeugenol (6.7 mM) to veratraldehyde, and α-methylstyrene (6.7 mM) to acetophenone by PsaPOX (1 U/mL) at RT. The presented product concentrations are the differences between the values determined for the reaction with the active and heat inactivated enzyme (blank) (the original values are shown in Table S4). (**b**) Decolorization of 7% (*v/v*) β-carotene and 7% (*v/v*) annatto by PsaPOX (1 U/L) at 40 °C. Cleavage of carotenoids was shown as extinction decrease per min. Values are the average of triplicate experiments with standard deviations shown as error bars.

In addition to the described substrates, PsaPOX (1 U/L) also showed an alkene cleavage activity towards the natural dyes β-carotene and annatto (mixture of the xanthophylls bixin and norbixin), which was detected by substrate bleaching (Figure 7b). The activity for annatto was higher than for β-carotene. Cleavage of β-carotene and annatto is also known for other fungal DyPs [19,39–41]. For example, cleavage of β-carotene by a DyP from *Lepista irina* resulted in formation of the volatiles β-ionone, β-cyclocitral, dihydroactinidiolide, and 2-hydroy-2,6,6-trimethylcyclohexanone [41].

3. Materials and Methods

3.1. Chemicals and Materials

Chemicals were obtained from Sigma-Aldrich (Seelze, Germany), Carl-Roth (Karlsruhe, Germany), or Merck (Darmstadt, Germany) in *P. a.* quality. Enzymes were from Thermo Fisher Scientific (Braunschweig, Germany), if not stated otherwise. PCR primers were obtained from Eurofins MWG Operon (Ebersberg, Germany).

3.2. Cultivation of *P. sapidus*

P. sapidus (Deutsche Sammlung von Mikroorganismen und Zellkulturen GmbH, DSMZ, strain no. 2866) was pre-grown on 1.5% (*w*/*v*) agar plates with standard nutrient liquid (SNL) medium and maintained at 4 °C until use [39]. For pre-cultivation, 1 cm^2 of grown agar was transferred to 100 mL SNL medium and homogenized using an Ultraturrax homogenizer (ART Prozess- & Labortechnik, Müllheim, Germany). The pre-cultures were incubated for 5 days at 150 rpm and 24 °C. Afterwards, 6.5 g of pre-grown mycelium was used to inoculate 250 mL SNL. The main culture was incubated at 150 rpm and 24 °C. After six days, the mycelium was separated from the culture supernatant by centrifugation (5000× *g*, 4 °C, 15 min) and lyophilized as described elsewhere [42].

3.3. Purification Strategy

Ten g of lyophilized mycelium were re-suspended in 400 mL buffer A (50 mM Bis-Tris, pH 6.0, 1 M $(NH_4)_2SO_4$) and extracted for 1 h at 4 °C in horizontal position in an orbital shaker. Insoluble components were removed by centrifugation (5000× *g*, 4 °C, 15 min) followed by filtration (PES filter,

0.45 μm, Merck). Subsequently, 80 mL filtered supernatant were applied on a Phenyl Sepharose fast flow column (20 mL, GE Healthcare Bio-Sciences AB, Uppsala, Sweden) pre-equilibrated with buffer A. After the column was washed with buffer A, the active enzyme was eluted with a linear gradient (130 mL, 100–0% buffer A) with 100% distilled water at a constant flow rate of 2 mL/min. Active fractions were pooled, desalted and concentrated by ultrafiltration (3 kDa cut off, polyethersulfone (PES), Sartorius, Göttingen, Germany). Concentrate (20 mL) was diluted two times with 20 mM sodium acetate pH 4.0 (buffer B) and loaded onto three linked HiTrap SP Sepharose columns (1 mL, GE Healthcare Bio-Sciences AB) pre-equilibrated with buffer B. Proteins were eluted with a stepwise ionic strength gradient (0, 20, 100% buffer C: 20 mM sodium acetate pH 4.0, 1 M NaCl) with 100% buffer C at a constant flow rate of 1 mL/min.

3.4. Gel Electrophoresis

SDS-PAGE analysis was performed as described elsewhere [43]. Semi-native PAGE was performed under non-denaturing conditions using 12% gels. For this, samples were prepared with a native loading buffer (without DTT and without 2% (*w*/*v*; 6.9 mM) SDS) and gel electrophoresis was performed at 10 mA per gel and 4 °C. Gels were stained with 0.5 mM ABTS (dissolved in 100 mM sodium acetate buffer pH 3.5 or 4.5) in the presence of 100 μM hydrogen peroxide for detection of peroxidases. For deglycosylation, samples were treated with 1 μL (500 U) endoglycosidase H (EndoH, New England BioLabs, Ipswich, MA, USA) in 20 μL for 2 h at 37 °C before gel electrophoresis.

3.5. Isoelectric Focussing

Analytical isoelectric focusing polyacrylamide gel electrophoresis was performed on a HPE^{TM} BlueHorizon^{TM} system (Serva Electrophoresis GmbH, Heidelberg, Germany) using Servalyt^{TM} Precotes^{TM} Precast Gels (Serva Electrophoresis GmbH) with an immobilized pH gradient of pH 3 to 10. To determine the isoelectric points of the enzymes, marker proteins (IEF-Marker 3-10, Serva Electrophoresis GmbH) were used. Gels were stained with Coomassie Brilliant Blue, or for specific visualization of peroxidases, with 1% (*w*/*v*; 9.2 mM) phenylendiamine and 1% (*w*/*v*; 10.6 mM) urea peroxide (dissolved in 100 mM sodium acetate buffer pH 3.5) at RT until a yellow band was observed.

3.6. Peptide Mass Fingerprinting

Protein bands were excised from SDS gels, dried, and tryptically hydrolysed. The resulting peptides were extracted and purified according to standard protocols and the amino acid sequence was analyzed with electrospray ionization-tandem mass spectrometry (ESI-MS/MS) using a maXis quadrupole time of flight (QTOF) mass spectrometer (Bruker, Bremen, Germany) as described previously [43,44]. The obtained partial sequences of PsaPOX and of the oxidase were used for a similarity search against public databases (NCBI BlastP).

3.7. cDNA Synthesis and Gene Amplification

Isolation of total RNA from mycelium of *P. sapidus* at culture day six and cDNA synthesis were performed as described previously [13] using the primer 5′-AAGCAGTGGTATCAACGCAGAGT ACGCTTTTTTTTTTTTTTTTTTTT-3′ for reverse transcription. Specific primers for gene amplification were deduced from the ORF-start (P1: 5′-ATGACTACACCTGCACCACCCCTCGACCTC-3′) and -stop (P2: 5′-TCAAGCAGAGATTGGAGCTTGGGTSWGAGGA-3′) region of the homologous peroxidase of *Pleurotus ostreatus* PC15 (GenBank accession no. KDQ22873.1). PCRs were performed with Phusion High-Fidelity DNA Polymerase and the Master Cycler gradient (Eppendorf, Hamburg, Germany) as described elsewhere [45]. The cycler program was as follows: denaturation for 2 min at 98 °C, 35 cycles at 98 °C for 1 min, 62 °C for 30 s and 72 °C for 90 s, and a final elongation at 72 °C for 10 min. Analysis of PCR products, ligation, transformation in *Escherichia coli*, colony PCR, and sequencing were performed as described by Behrens et al. [13]. Translation of DNA sequences was performed using SnapGene® (GSL Biotech LLC, Chicago, IL, USA). Sequence homology was examined using BLAST [46]. Alignments were produced by ClustalOmega [47].

3.8. Heterologous Expression of PsaPOX in Komagataella pfaffii

The gene of PsaPOX was amplified with a C-terminal 6x His tag using the primers PsaPOX_fw 5′-AAAAAGAATTCatgactacacctgcaccacccctcgacctcaacaa-3′ and PsaPOX_rev 5′-atatatGCGGCCGC tcaGTGGTGATGGTGATGATGggtagagatcggagcctgggcctg-3′ (underlined are the EcoRI and NotI restriction sites, respectively; lower cases represent parts of the coding *PsaPOX*). In addition, it was inserted in frame with the *Saccharomyces cerevisiae* α-factor secretion signal sequence into the *K. pfaffii* pPIC9 expression vector (Invitrogen, Karlsruhe, Germany). The resulting expression construct pPIC-PsaPOX-His was transformed into *E. coli* TOP10 for vector propagation, isolated (NucleoSpin, Macherey-Nagel, Düren, Germany), linearized with PmeI, and used for transformation of *K. pfaffii* GS115 according to a standard protocol [48]. The linearized empty vector was transformed in the same way and served as negative control. Forty-eight transformants were tested for peroxidase activity after selection according to their ability to grow on histidine-deficient agar plates in 96-well plates for 120 h, as described elsewhere [49]. Gene expression was induced by daily addition of 1% (*v*/*v*) methanol.

3.9. His-Tag Purification of Recombinant PsaPOX

For purification of the His-tag labelled recombinant enzyme from *K. pfaffii* culture supernatant, Ni-NTA affinity chromatography was used according to Nieter et al. [50].

3.10. Biotransformation

Transformation of *trans*-anethole was carried out in 4 mL gas tight glass vials in horizontal position at a shaking rate of 200 rpm for 16 h at RT in the absence of light. Reaction mixtures contained 30 mg *P. sapidus* lyophilisate or 100 μL liquid sample buffered in Bis-Tris (50 mM, pH 6) with or without addition of 1 mM manganese sulfate in a total volume of 1 mL and 1 μL (6.7 mM) *trans*-anethole. Blanks (chemical: without lyophilisate or liquid sample; biological: with heat inactivated mycelium (1 h at 95 °C)) were performed the same way. All experiments were performed as duplicates. After incubation, *trans*-anethole and its conversion product *p*-anisaldehyde were extracted with 1 mL hexane containing 100 mg/L (1 mM) cyclohexanol as internal standard (IS). The organic phase was dried with anhydrous sodium sulfate and subsequently analyzed by gas chromatography (GC). GC measurements were performed with an Agilent 7890 instrument equipped with a DB-WAX UI column (30 m × 0.32 mm, 0.25 μm, Agilent, Santa Clara, CA, USA), a split/splitless injector port (1:5) and a flame ionization detection (FID) system. Hydrogen was used as carrier gas at a constant flow rate of 2.1 mL per minute. One μL sample was injected via an autosampler and measured using the following method: 40 °C (3 min), a temperature increase of 10 °C per minute until 230 °C and a final hold time of 10 min. The *trans*-anethole and *p*-anisaldehyde were semi-quantified referring to the area of the internal standard. Biotransformation products were identified using standards and comparison of retention indices with literature.

3.11. Enzyme Activities

Total peroxidase activity was determined photometrically (EON™ High Performance Microplate Spectrophotometer, BioTek Instruments GmbH, Bad Friedrichshall, Germany) by monitoring the oxidation of ABTS in the presence of hydrogen peroxide at 420 nm ($\varepsilon_{420} = 3.6 \times 10^4$ M^{-1} cm^{-1}) and 30 °C for 10 min. For this, the samples were mixed with sodium acetate buffer (100 mM, pH 4.0 or pH 3.5), 0.1 mM hydrogen peroxide, and 0.5 mM ABTS in a total volume of 300 μL. One unit of enzyme activity was defined as 1 μmol substrate oxidized per minute under the experimental conditions.

To determine manganese peroxidase activity, samples were mixed with manganese sulfate (1 mM), malonate buffer (100 mM, pH 3.5), and hydrogen peroxide (0.1 mM) in a total volume of 300 μL.

Mn^{3+} formation was monitored photometrically at 270 nm (ε_{270} = 1.16 × 10^4 M^{-1} cm^{-1}) and 30 °C for 30 min. One unit of enzyme activity was defined as 1 µmol Mn^{3+} per minute released by manganese peroxidases at the given conditions.

Decolorization of Reactive blue 19 (RB19, 150 µM) and Reactive black 5 (RB5, 80 µM) by PsaPOX (1 U/L; 0.25 mg/L) was tested. The respective anthraquinone dye and the enzyme was incubated in the presence of 100 µM hydrogen peroxide, 25 mM manganese sulfate, and 100 mM sodium acetate buffer pH 3.5 in a total volume of 300 µL at 40 °C for 20 min. Decolorization was monitored photometrically at 595 nm (RB19; ε_{595} = 1.0 × 10^4 M^{-1} cm^{-1}) or 598 nm (RB5; ε_{598} =3.0 × 10^4 M^{-1} cm^{-1}). One unit of enzyme activity was defined as 1 µmol dye degraded per minute at the given conditions.

All enzyme assays were performed as triplicates. Blanks were carried out with water instead of enzyme and by omission of hydrogen peroxide.

3.12. Biochemical Characterization of PsaPOX

Effects of pH and temperature on peroxidase activity of PsaPOX (0.25 mg/L) were analyzed with ABTS as substrate as described above (see "4.11"). Relative activities were normalized to the highest activity and residual activities to the initial activity prior incubation. The pH optimum was determined using Britton-Robinson buffer [51] in a range of pH 2.0–9.5 instead of sodium acetate buffer. For determination of the temperature optimum the activity assay was performed at different temperatures (20–90 °C) at pH 3.5, whilst for analysis of the temperature stability the enzyme was incubated for 1 h at 20–90 °C prior enzyme activity measurement at pH 3.5 and 40 °C. For the analysis of pH-stability PsaPOX was incubated in Britton-Robinson buffer from pH 2.0 to 9.5 for 1 h at RT before the peroxidase activity was examined at pH 3.5 and 40 °C.

Hydrogen peroxide as well as Mn^{2+} dependency of PsaPOX were determined for PsaPOX by evaluation of peroxidase activity as described above ("4.11") with changing hydrogen peroxide and manganese sulfate concentrations (H_2O_2: 0–1 mM H_2O_2, without addition of $MnSO_4$; Mn^{2+}: 100 µM H_2O_2, 0–100 mM $MnSO_4$) at optimal pH and thermal conditions. Kinetic constants of PsaPOX were calculated for Mn^{2+} and ABTS (0–300 µM ABTS in the presence of 100 µM H_2O_2 and 25 mM $MnSO_4$) by SigmaPlot 12.5 (Systat Software Inc., Chicago, IL, USA) with nonlinear regression. Protein concentrations were determined according to Lowry et al. [52] using bovine serum albumin as standard.

3.13. Alkene Cleavage Activity of PsaPOX

The purified recombinant PsaPOX was used for transformation of *trans*-anethole as mentioned above ("4.10") to confirm alkene cleavage activity. For this, 1 U/mL (0.25 mg/mL) of enzyme was used for biotransformation. Biotransformation was performed with 100 mM sodium acetate buffer pH 3.5 in the presence of 100 µM hydrogen peroxide and 25 mM manganese sulfate at RT for 16 h. Biotransformation of the alkenes methyl isoeugenol (6.7 mM), α-methylstyrene (6.7 mM), and piperine (0.7 mM) was tested accordingly. Blanks were performed without enzyme (chemical blank) or with heat inactivated enzyme (1 h at 95 °C, biological blank). The determined product concentrations for the blanks were subtracted from the concentrations yielded for the reaction with the active enzyme to calculate the enzymatically generated product concentration. For carotene degradation, a β-carotene emulsion was prepared according to Linke et al. [43]. 7% (*v*/*v*) of β-carotene emulsion or annatto (Chr. Hansen, Nienburg, Germany, Prod. No. 240569), 100 µM hydrogen peroxide, 25 mM manganese sulfate, 100 mM sodium acetate pH 3.5, and 1 U/L (0.25 mg/L) PsaPOX in a total volume of 300 µL was incubated at 40 °C for 20 min. Alkene cleavage of both substrates was measured photometrically as extinction decrease at 455 nm.

3.14. Detection of Hydrogen Peroxide

For the detection of H_2O_2, 75 µL IEX fraction, 50 mM Bis-Tris pH 6.0, 6.7 mM *trans*-anethole, 10 U/mL HRP (Sigma Aldrich), and 0.5 mM *o*-dianisidine in a total volume of 300 µL were incubated at

RT for 1 h. In the presence of hydrogen peroxide, formation of a red-brown reaction product occurred. Blanks were performed with 50 mM Bis-Tris pH 6.0 instead of IEX fraction.

3.15. Sequence Accession Numbers

The nucleotide sequence of the *PsaPOX* gene has been deposited in the GenBank database under accession number MT043310.

4. Conclusions

A DyP-type peroxidase of *P. sapidus* with alkene cleavage activity as well as the corresponding gene were identified and the gene was heterologously expressed in *Komagataella pfaffii*. The PsaPOX possessed typical sequence motifs, structural topology, and catalytic residues as described for DyPs, even though the decolorization of the anthraquinone Reactive blue 19, a common reaction for DyPs, was not observed. A non-canonical Mn^{2+}-oxidation site on the protein surface was detected, which allows PsaPOX to oxidize Mn^{2+}. After biochemical characterization, the alkene cleavage activity of PsaPOX towards different aryl alkenes was confirmed by biotransformation. PsaPOX is the first described DyP-type peroxidase with such an activity. In addition, bleaching of β-carotene and annatto was determined. The results for the alkene cleavage underline the potential of the PsaPOX as biocatalyst for the generation of aromatic aldehydes with olfactory properties, such as *p*-anisaldehyde, veratraldehyde, or acetophenone, which are used in the fragrance and flavor industry [1]. Improvement of the conversions and product yields may be accomplished by protein engineering, as has been shown for the alkene cleaving manganese-dependent Cupin TM1459 from *Thermotoga maritima* [53]. Another application beyond aroma production could be carotene bleaching of whey or wheat dough.

Supplementary Materials: The following are available online, Figure S1: Structural homology model of PsaPOX, Figure S2: Detection of hydrogen peroxide in the IEX fraction with *o*-dianisidine and HRP in the presence of *trans*-anethole after 1 h of incubation at pH 6.0 and RT, Figure S3: Alignment of the hypothetical protein (KDQ29984.1) from *P. ostreatus*, which was the best hit for the 75 kDa band of the IEX fraction by a homology search against the public database NCBI, and other members of the GMC oxidoreductase family, Figure S4: Stability of PsaPOX during biotransformation of *trans*-anethole over 16 h, Table S1: *p*-Anisaldehyde concentration after biotransformation of *trans*-anethole with different basidiomycetes, Table S2: *p*-Anisaldehyde concentration after biotransformation of *trans*-anethole with the active IEX fraction in the presence or absence of Mn^{2+} and/or H_2O_2 for 16 h at RT, Table S3: *p*-Anisaldehyde concentration after bioconversion of *trans*-anethole by recombinant PsaPOX with and without addition of H_2O_2 and Mn^{2+} for 16 h at RT, Table S4: *p*-Anisaldehyde concentration after biotransformation of *trans*-anethole, (*E*)-methyl isoeugenol, and α-methylstyrene by recombinant PsaPOX (1 U/mL) in the presence of 100 μM H_2O_2 and 25 mM $MnSO_4$ for 16 h at pH 3.5 and RT.

Author Contributions: Conceptualization, N.-K.K. and R.G.B.; methodology, N.-K.K.; validation, N.-K.K., R.G.B. and F.E.; formal analysis, N.-K.K.; investigation, N.K.K; writing—original draft preparation, N.-K.K.; writing—review and editing, R.G.B. and F.E.; visualization, N.-K.K.; supervision, F.E.; project administration, R.G.B.; funding acquisition, R.G.B. All authors have read and agreed to the published version of the manuscript.

Funding: This research was funded the BMBF cluster Bioeconomy International 2015, grant number 031B0307A. The APC was funded by the Open Access fund of the Gottfried Wilhelm Leibniz Universität Hannover.

Acknowledgments: B. Fuchs and A. Nieter are thanked for detecting the cleavage reaction and the peroxidase activity during the screenings.

Conflicts of Interest: The authors declare no conflict of interest.

References

1. Fahlbusch, K.-G.; Hammerschmidt, F.-J.; Panten, J.; Pickenhagen, W.; Schatkowski, D.; Bauer, K.; Garbe, D.; Surburg, H. Flavors and Fragrances. In *Ullmann's Encyclopedia of Industrial Chemistry*; Wiley-VCH Verlag GmbH & Co. KGaA: Weinheim, Germany, 2003; pp. 73–140.
2. Spannring, P.; Bruijnincx, P.C.A.; Weckhuysen, B.M.; Klein Gebbink, R.J.M. Transition metal-catalyzed oxidative double bond cleavage of simple and bio-derived alkenes and unsaturated fatty acids. *Catal. Sci. Technol.* **2014**, *4*, 2182–2209. [CrossRef]

3. Rajagopalan, A.; Lara, M.; Kroutil, W. Oxidative Alkene Cleavage by Chemical and Enzymatic Methods. *Adv. Synth. Catal.* **2013**, *355*, 3321–3335. [CrossRef]
4. Criegee, R. Mechanism of Ozonolysis. *Angew. Chemie Int. Ed. Engl.* **1975**, *14*, 745–752. [CrossRef]
5. Schrader, J.; Etschmann, M.M.W.; Sell, D.; Hilmer, J.-M.; Rabenhorst, J. Applied biocatalysis for the synthesis of natural flavour compounds – current industrial processes and future prospects. *Biotechnol. Lett.* **2004**, *26*, 463–472. [CrossRef] [PubMed]
6. Rajagopalan, A.; Schober, M.; Emmerstorfer, A.; Hammerer, L.; Migglautsch, A.; Seisser, B.; Glueck, S.M.; Niehaus, F.; Eck, J.; Pichler, H.; et al. Enzymatic Aerobic Alkene Cleavage Catalyzed by a Mn^{3+}-Dependent Proteinase A Homologue. *ChemBioChem* **2013**, *14*, 2427–2430. [CrossRef] [PubMed]
7. Han, D.; Ryu, J.-Y.; Kanaly, R.A.; Hur, H.-G. Isolation of a Gene Responsible for the Oxidation of *trans*-Anethole to *para*-Anisaldehyde by *Pseudomonas putida* JYR-1 and Its Expression in *Escherichia coli*. *Appl. Environ. Microbiol.* **2012**, *78*, 5238–5246. [CrossRef] [PubMed]
8. Yamada, M.; Okada, Y.; Yoshida, T.; Nagasawa, T. Purification, characterization and gene cloning of isoeugenol-degrading enzyme from *Pseudomonas putida* IE27. *Arch. Microbiol.* **2007**, *187*, 511–517. [CrossRef]
9. Hajnal, I.; Faber, K.; Schwab, H.; Hall, M.; Steiner, K. Oxidative Alkene Cleavage Catalysed by Manganese-Dependent Cupin TM1459 from *Thermotoga maritima*. *Adv. Synth. Catal.* **2015**, *357*, 3309–3316. [CrossRef]
10. Tuynman, A.; Spelberg, J.L.; Kooter, I.M.; Schoemaker, H.E.; Wever, R. Enantioselective Epoxidation and Carbon–Carbon Bond Cleavage Catalyzed by *Coprinus cinereus* Peroxidase and Myeloperoxidase. *J. Biol. Chem.* **2000**, *275*, 3025–3030. [CrossRef]
11. Mutti, F.G.; Lara, M.; Kroutil, M.; Kroutil, W. Ostensible Enzyme Promiscuity: Alkene Cleavage by Peroxidases. *Chem. Eur. J.* **2010**, *16*, 14142–14148. [CrossRef]
12. Hofrichter, M.; Ullrich, R.; Pecyna, M.J.; Liers, C.; Lundell, T. New and classic families of secreted fungal heme peroxidases. *Appl. Microbiol. Biotechnol.* **2010**, *87*, 871–897. [CrossRef] [PubMed]
13. Behrens, C.J.; Zelena, K.; Berger, R.G. Comparative Cold Shock Expression and Characterization of Fungal Dye-Decolorizing Peroxidases. *Appl. Biochem. Biotechnol.* **2016**, *179*, 1404–1417. [CrossRef] [PubMed]
14. Sugano, Y. DyP-type peroxidases comprise a novel heme peroxidase family. *Cell. Mol. Life Sci.* **2009**, *66*, 1387–1403. [CrossRef] [PubMed]
15. Linde, D.; Ruiz-Dueñas, F.J.; Fernández-Fueyo, E.; Guallar, V.; Hammel, K.E.; Pogni, R.; Martínez, A.T. Basidiomycete DyPs: Genomic diversity, structural–functional aspects, reaction mechanism and environmental significance. *Arch. Biochem. Biophys.* **2015**, *574*, 66–74. [CrossRef]
16. Strittmatter, E.; Liers, C.; Ullrich, R.; Wachter, S.; Hofrichter, M.; Plattner, D.A.; Piontek, K. First Crystal Structure of a Fungal High-redox Potential Dye-decolorizing Peroxidase. *J. Biol. Chem.* **2013**, *288*, 4095–4102. [CrossRef]
17. Rais, D.; Zibek, S. Biotechnological and Biochemical Utilization of Lignin. In *Biorefineries*; Wagemann, K., Tippkötter, N., Eds.; Springer International Publishing: Cham, Switzerland, 2017; pp. 469–518.
18. Salvachua, D.; Prieto, A.; Martinez, A.T.; Martinez, M.J. Characterization of a Novel Dye-Decolorizing Peroxidase (DyP)-Type Enzyme from *Irpex lacteus* and Its Application in Enzymatic Hydrolysis of Wheat Straw. *Appl. Environ. Microbiol.* **2013**, *79*, 4316–4324. [CrossRef]
19. Lauber, C.; Schwarz, T.; Nguyen, Q.K.; Lorenz, P.; Lochnit, G.; Zorn, H. Identification, heterologous expression and characterization of a dye-decolorizing peroxidase of *Pleurotus sapidus*. *AMB Express* **2017**, *7*, 164. [CrossRef]
20. Busse, N.; Wagner, D.; Kraume, M.; Czermak, P. Reaction Kinetics of Versatile Peroxidase for the Degradation of Lignin Compounds. *Am. J. Biochem. Biotechnol.* **2013**, *9*, 365–394. [CrossRef]
21. Arnao, M.B.; Acosta, M.; del Rio, J.A.; Varón, R.; García-Cánovas, F. A kinetic study on the suicide inactivation of peroxidase by hydrogen peroxide. *Biochim. Biophys. Acta Protein Struct. Mol. Enzymol.* **1990**, *1041*, 43–47. [CrossRef]
22. Dako, E.; Bernier, A.-M.; Thomas, A.; Jankowski, C.K. The Problems Associated with Enzyme Purification. In *Chemical Biology*; Ekinci, D., Ed.; InTech: Rijeka, Croatia, 2012; pp. 19–40.
23. Schulz, K.; Nieter, A.; Scheu, A.-K.; Copa-Patiño, J.L.; Thiesing, D.; Popper, L.; Berger, R.G. A type D ferulic acid esterase from *Streptomyces werraensis* affects the volume of wheat dough pastries. *Appl. Microbiol. Biotechnol.* **2018**, *102*, 1269–1279. [CrossRef]

24. Kolwek, J.; Behrens, C.; Linke, D.; Krings, U.; Berger, R.G. Cell-free one-pot conversion of (+)-valencene to (+)-nootkatone by a unique dye-decolorizing peroxidase combined with a laccase from *Funalia trogii*. *J. Ind. Microbiol. Biotechnol.* **2018**, *45*, 89–101. [CrossRef]
25. Hofrichter, M. Review: Lignin conversion by manganese peroxidase (MnP). *Enzyme Microb. Technol.* **2002**, *30*, 454–466. [CrossRef]
26. Nakayama, T.; Amachi, T. Fungal peroxidase: Its structure, function, and application. *J. Mol. Catal. B Enzym.* **1999**, *6*, 185–198. [CrossRef]
27. Fernández-Fueyo, E.; Linde, D.; Almendral, D.; López-Lucendo, M.F.; Ruiz-Dueñas, F.J.; Martínez, A.T. Description of the first fungal dye-decolorizing peroxidase oxidizing manganese(II). *Appl. Microbiol. Biotechnol.* **2015**, *99*, 8927–8942. [CrossRef] [PubMed]
28. Fernández-Fueyo, E.; Davó-Siguero, I.; Almendral, D.; Linde, D.; Baratto, M.C.; Pogni, R.; Romero, A.; Guallar, V.; Martínez, A.T. Description of a Non-Canonical Mn(II)-Oxidation Site in Peroxidases. *ACS Catal.* **2018**, *8*, 8386–8395. [CrossRef]
29. Linde, D.; Pogni, R.; Cañellas, M.; Lucas, F.; Guallar, V.; Baratto, M.C.; Sinicropi, A.; Sáez-Jiménez, V.; Coscolín, C.; Romero, A.; et al. Catalytic surface radical in dye-decolorizing peroxidase: a computational, spectroscopic and site-directed mutagenesis study. *Biochem. J.* **2015**, *466*, 253–262. [CrossRef]
30. Amara, S.; Perrot, T.; Navarro, D.; Deroy, A.; Benkhelfallah, A.; Chalak, A.; Daou, M.; Chevret, D.; Faulds, C.B.; Berrin, J.-G.; et al. Enzyme Activities of Two Recombinant Heme-Containing Peroxidases, Tv DyP1 and Tv VP2, Identified from the Secretome of *Trametes versicolor*. *Appl. Environ. Microbiol.* **2018**, *84*, e02826-17. [CrossRef]
31. Duan, Z.; Shen, R.; Liu, B.; Yao, M.; Jia, R. Comprehensive investigation of a dye-decolorizing peroxidase and a manganese peroxidase from *Irpex lacteus* F17, a lignin-degrading basidiomycete. *AMB Express* **2018**, *8*, 119. [CrossRef]
32. Passardi, F.; Theiler, G.; Zamocky, M.; Cosio, C.; Rouhier, N.; Teixera, F.; Margis-Pinheiro, M.; Ioannidis, V.; Penel, C.; Falquet, L.; et al. PeroxiBase: The peroxidase database. *Phytochemistry* **2007**, *68*, 1605–1611. [CrossRef]
33. Colpa, D.I.; Fraaije, M.W.; van Bloois, E. DyP-type peroxidases: a promising and versatile class of enzymes. *J. Ind. Microbiol. Biotechnol.* **2014**, *41*, 1–7. [CrossRef]
34. Romanos, M. Advances in the use of *Pichia pastoris* for high-level gene expression. *Curr. Opin. Biotechnol.* **1995**, *6*, 527–533. [CrossRef]
35. Ogawa, S.; Shimizu, T.; Ohki, H.; Araya, T.; Okuno, T.; Miyairi, K. Expression, purification, and analyses of glycosylation and disulfide bonds of *Stereum purpureum* endopolygalacturonase I in *Pichia pastoris*. *Protein Expr. Purif.* **2009**, *65*, 15–22. [CrossRef] [PubMed]
36. Behrens, C.J.; Linke, D.; Allister, A.B.; Zelena, K.; Berger, R.G. Variants of PpuLcc, a multi-dye decolorizing laccase from *Pleurotus pulmonarius* expressed in *Pichia pastoris*. *Protein Expr. Purif.* **2017**, *137*, 34–42. [CrossRef] [PubMed]
37. Avram, A.; Sengupta, A.; Pfromm, P.H.; Zorn, H.; Lorenz, P.; Schwarz, T.; Nguyen, K.Q.; Czermak, P. Novel DyP from the basidiomycete *Pleurotus sapidus*: substrate screening and kinetics. *Biocatalysis* **2018**, *4*, 1–13. [CrossRef]
38. Rahmanpour, R.; Bugg, T.D.H. Characterisation of Dyp-type peroxidases from *Pseudomonas fluorescens* Pf-5: Oxidation of Mn(II) and polymeric lignin by Dyp1B. *Arch. Biochem. Biophys.* **2015**, *574*, 93–98. [CrossRef]
39. Linke, D.; Leonhardt, R.; Eisele, N.; Petersen, L.M.; Riemer, S.; Nimtz, M.; Berger, R.G. Carotene-degrading activities from *Bjerkandera adusta* possess an application in detergent industries. *Bioprocess Biosyst. Eng.* **2015**, *38*, 1191–1199. [CrossRef]
40. Scheibner, M.; Hülsdau, B.; Zelena, K.; Nimtz, M.; de Boer, L.; Berger, R.G.; Zorn, H. Novel peroxidases of *Marasmius scorodonius* degrade β-carotene. *Appl. Microbiol. Biotechnol.* **2008**, *77*, 1241–1250. [CrossRef]
41. Zorn, H.; Langhoff, S.; Scheibner, M.; Nimtz, M.; Berger, R.G. A Peroxidase from *Lepista irina* Cleaves β, β-Carotene to Flavor Compounds. *Biol. Chem.* **2003**, *384*, 1049–1056. [CrossRef]
42. Krings, U.; Lehnert, N.; Fraatz, M.A.; Hardebusch, B.; Zorn, H.; Berger, R.G. Autoxidation versus biotransformation of α-pinene to flavors with *Pleurotus sapidus*: Regioselective hydroperoxidation of α-pinene and stereoselective dehydrogenation of verbenol. *J. Agric. Food Chem.* **2009**, *57*, 9944–9950. [CrossRef]

43. Linke, D.; Omarini, A.B.; Takenberg, M.; Kelle, S.; Berger, R.G. Long-Term Monokaryotic Cultures of *Pleurotus ostreatus* var. *florida* Produce High and Stable Laccase Activity Capable to Degrade ß-Carotene. *Appl. Biochem. Biotechnol.* **2019**, *187*, 894–912. [CrossRef]
44. Schulz, K.; Giesler, L.; Linke, D.; Berger, R.G. A prolyl endopeptidase from *Flammulina velutipes* for the possible degradation of celiac disease provoking toxic peptides in cereal proteins. *Process Biochem.* **2018**, *73*, 47–55. [CrossRef]
45. Nieter, A.; Haase-Aschoff, P.; Linke, D.; Nimtz, M.; Berger, R.G. A halotolerant type A feruloyl esterase from *Pleurotus eryngii*. *Fungal Biol.* **2014**, *118*, 348–357. [CrossRef] [PubMed]
46. Altschul, S.F.; Gish, W.; Miller, W.; Myers, E.W.; Lipman, D.J. Basic local alignment search tool. *J. Mol. Biol.* **1990**, *215*, 403–410. [CrossRef]
47. Sievers, F.; Wilm, A.; Dineen, D.; Gibson, T.J.; Karplus, K.; Li, W.; Lopez, R.; McWilliam, H.; Remmert, M.; Söding, J.; et al. Fast, scalable generation of high-quality protein multiple sequence alignments using Clustal Omega. *Mol. Syst. Biol.* **2011**, *7*, 539. [CrossRef] [PubMed]
48. Lin-Cereghino, J.; Wong, W.W.; Xiong, S.; Giang, W.; Luong, L.T.; Vu, J.; Johnson, S.D.; Lin-Cereghino, G.P. Condensed protocol for competent cell preparation and transformation of the methylotrophic yeast *Pichia pastoris*. *Biotechniques* **2005**, *38*, 44–48. [CrossRef]
49. Sygmund, C.; Gutmann, A.; Krondorfer, I.; Kujawa, M.; Glieder, A.; Pscheidt, B.; Haltrich, D.; Peterbauer, C.; Kittl, R. Simple and efficient expression of *Agaricus meleagris* pyranose dehydrogenase in *Pichia pastoris*. *Appl. Microbiol. Biotechnol.* **2012**, *94*, 695–704. [CrossRef]
50. Nieter, A.; Kelle, S.; Takenberg, M.; Linke, D.; Bunzel, M.; Popper, L.; Berger, R.G. Heterologous production and characterization of a chlorogenic acid esterase from *Ustilago maydis* with a potential use in baking. *Food Chem.* **2016**, *209*, 1–9. [CrossRef]
51. Britton, H.T.S.; Robinson, R.A. CXCVIII.—Universal buffer solutions and the dissociation constant of veronal. *J. Chem. Soc.* **1931**, 1456–1462. [CrossRef]
52. Lowry, O.H.; Rosebrpugh, N.J.; Randall, R.J. Protein measurement with the Folin phenol reagent. *J. Biol. Chem.* **1951**, *193*, 265–275.
53. Fink, M.; Trunk, S.; Hall, M.; Schwab, H.; Steiner, K. Engineering of TM1459 from *Thermotoga maritima* for increased oxidative alkene cleavage activity. *Front. Microbiol.* **2016**, *7*, 1511. [CrossRef]

Sample Availability: Samples of the pPIC9 vector containing the *PsaPOX* gene are available from the authors.

Article

Enrichment of Polyglucosylated Isoflavones from Soybean Isoflavone Aglycones Using Optimized Amylosucrase Transglycosylation

Young Sung Jung [1,†], Ye-Jin Kim [2,†], Aaron Taehwan Kim [1], Davin Jang [2], Mi-Seon Kim [2], Dong-Ho Seo [3], Tae Gyu Nam [4], Chan-Su Rha [1], Cheon-Seok Park [2] and Dae-Ok Kim [1,2,*]

1 Department of Food Science and Biotechnology, Kyung Hee University, Yongin 17104, Korea; chembio@khu.ac.kr (Y.S.J.); axk369@khu.ac.kr (A.T.K.); chansurha@khu.ac.kr (C.-S.R.)

2 Graduate School of Biotechnology, Kyung Hee University, Yongin 17104, Korea; jinatiger@khu.ac.kr (Y.-J.K.); davin1031@khu.ac.kr (D.J.); miseonkim95@khu.ac.kr (M.-S.K.); cspark@khu.ac.kr (C.-S.P.)

3 Department of Food Science and Technology, Jeonbuk National University, Jeonju 54896, Korea; dhseo@jbnu.ac.kr

4 Food Analysis Center, Korea Food Research Institute, Wanju 55365, Korea; ntg97@kfri.re.kr

* Correspondence: DOKIM05@khu.ac.kr; Tel.: +82-31-201-3796

† These authors contributed equally to this work.

Academic Editor: Josefina Aleu

Received: 6 December 2019; Accepted: 30 December 2019; Published: 1 January 2020

Abstract: Isoflavones in soybeans are well-known phytoestrogens. Soy isoflavones present in conjugated forms are converted to aglycone forms during processing and storage. Isoflavone aglycones (IFAs) of soybeans in human diets have poor solubility in water, resulting in low bioavailability and bioactivity. Enzyme-mediated glycosylation is an efficient and environmentally friendly way to modify the physicochemical properties of soy IFAs. In this study, we determined the optimal reaction conditions for *Deinococcus geothermalis* amylosucrase-mediated α-1,4 glycosylation of IFA-rich soybean extract to improve the bioaccessibility of IFAs. The conversion yields of soy IFAs were in decreasing order as follows: genistein > daidzein > glycitein. An enzyme quantity of 5 U and donor:acceptor ratios of 1000:1 (glycitein) and 400:1 (daidzein and genistein) resulted in high conversion yield (average 95.7%). These optimal reaction conditions for transglycosylation can be used to obtain transglycosylated IFA-rich functional ingredients from soybeans.

Keywords: *Deinococcus geothermalis*; glycosyltransferase; *Glycine max* (L.) Merr.; HPLC/MS; isoflavone aglycone-rich extract; isoflavone α-glucoside

---

1. Introduction

Globally, soybean (*Glycine max* (L.) Merr.) is a large portion of major grain production, along with corn, wheat, and rice [1]. Isoflavones, a subgroup of flavonoids, are well-known plant secondary metabolites that are found mainly in the plant families *Fabaceae* or *Leguminosae*, including soybean [2,3]. Isoflavones in soybeans are present mainly in conjugated forms [4]. Soy isoflavones are classified as acetylisoflavones, malonylisoflavones, non-acylated isoflavone glucosides, and isoflavone aglycones (IFAs), depending on the side groups attached to the isoflavone skeleton [5]. IFAs have antioxidant properties, interact with epigenetic modifications, and exhibit estrogen-like activities [3]. Dietary IFAs can be metabolized to equol (4′,7-isoflavandiol), which has significant estrogenic activity in the intestine [6]. Soy isoflavones exist naturally in conjugated forms rather than as aglycones; for example, daidzein [7-hydroxy-3-(4-hydroxyphenyl)chromen-4-one], glycitein [7-hydroxy-3-(4-hydroxyphenyl)-6-methoxychromen-4-one], and genistein [5,7-dihydroxy-3-(4-hydroxyphenyl)chromen-4-one] [5]. However, conjugated isoflavones are converted into aglycone forms during processing or storage [4]. The bioaccessibility of IFAs is low due to their poor solubility in water [7]. Therefore, various methods have been developed to overcome these shortcomings of IFAs [8,9].

Glycosylation via chemical or non-chemical methods can change the physicochemical properties of compounds [10,11]. The use of enzymes, such as glycosyltransferases, is a natural way to improve the solubility of IFA [12]. Enzymatic glycosylation reactions are efficient and environmentally friendly [13]. Compared with harsh chemical methods, enzymatic methods typically generate five times less waste and have a 15-fold higher space–time yield [14]. Recently, biological glycosylation methods using a variety of enzymes to improve the water solubility of lipophilic compounds have been developed [11,15,16]. Transglycosylation improved genistein solubility by 3700–44,000 [17] and transglycosylated catechin, naringin, and rutin were found to have 100-, 1000-, and 30,000-fold higher water solubility than non-transglycosylated catechin, naringin, and rutin, respectively [18–20]. The use of glycosyltransferases from microbial sources for transglycosylation reactions is an efficient and eco-friendly biological alternative to chemical reactions [21]. Flavonoids can be glycosylated using enzymes from microbial sources such as cyclodextrin glycosyltransferase (CGTase) and uridine diphosphate glucose (UDP-glucose) [16,22–24]. CGTase is widely employed to increase the solubility and stability of polyphenols such as glycosylated isoflavones [23,25]. However, the CGTase-reacted complex solution is cloudy and contains high molecular weight compounds [26], while the UDP-glucose complex has low reaction yields and is expensive [16,27].

Amylosucrase (EC 2.4.1.4; AS) is a glycosyltransferase enzyme that belongs to the glucoside hydrolase family 13 [28]. One of the main characteristics of AS is that it catalyzes transglycosylation reactions to create α-1,4-glycosidic linkages given sucrose as a substrate [28,29]; AS promotes sucrose hydrolysis to release glucose and fructose, and α-1,4-oligosaccharides are formed using the released glucose as an acceptor [30]. *Deinococcus geothermalis* AS (DGAS) synthesizes α-1,4-glycosidic bonds using various acceptors such as flavonoids [31]. DGAS enhances the water solubility of flavonoid aglycones by transglycosylation [21,31,32]. Compared with glycosyltransferase enzymes such as CGTase, DGAS can react with high substrate specificity using relatively inexpensive donors [27]. In addition, the interpretation of experimental results is easier when using DGAS because of fewer impurities and the limited number of transglycosylated sugar moieties [27,32,33]. These features of DGAS can be exploited to increase the solubility and bioavailability of bioactive phenolic compounds in soybeans.

In this study, we investigated the physicochemical changes of major soybean IFAs in response to DGAS-mediated transglycosylation. To determine what reaction conditions resulted in high transglycosylated conversion yields of IFAs, factors such as the amount of donors, acceptors, and enzymes were evaluated. Optimal reaction conditions were applied to soybean isoflavone extract to confirm that the IFAs were efficiently transglycosylated. Based on our results, we propose an application of the DGAS enzyme process to enrich transglycosylated IFAs in soy-based foods such as soybeans, soybean extracts, and soy products in the industrial field.

2. Results and Discussion

2.1. DGAS Expression and Activities

DGAS was successfully expressed by *Escherichia coli* transformed with the shuttle vector pHCXHD-DGAS as confirmed by the detection of *dgas* expression in transformed cells (Figure 1a). Activity of AS was observed in cell extracts of pHCXHD-DGAS-transformed cells. DGAS fused with 6× histidine appeared as a single band with a molecular mass of about 73 kDa (Figure 1b), which is in good agreement with the estimated molecular mass of DGAS. This result confirmed that DGAS was successfully expressed as a functional protein in *E. coli*. DGAS was considered to be free of protein toxicity because it was not harmful to *E. coli*. Temperature and pH profiles of DGAS expressed in *E. coli* were investigated in the range of 30 to 55 °C and pH 4.0 to 9.0, respectively, based on a previous report of DGAS expression in *E. coli* [30]. It was previously reported that isoflavone glucosides can be thermally transformed to their corresponding aglycone forms during processing [34]. In consideration of the stability of the IFAs, the temperature and pH for the DGAS-mediated transglycosylation of IFAs in this study were set at 45 °C and pH 5.0, respectively.

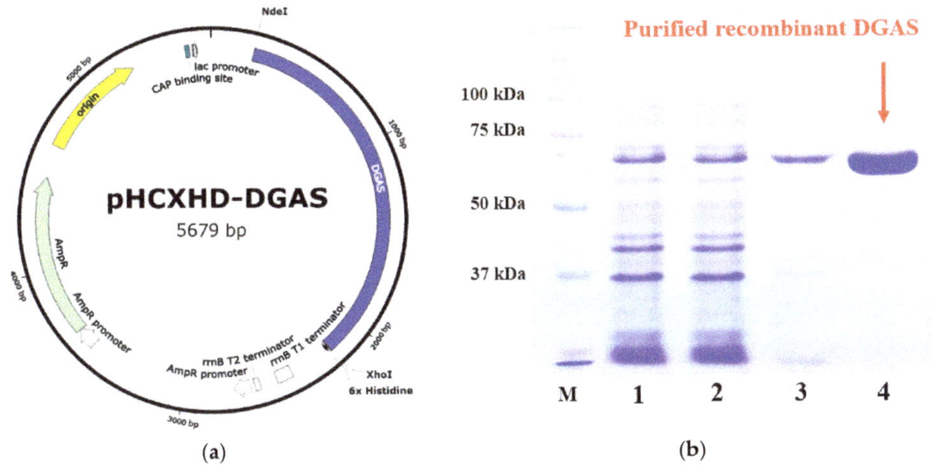

Figure 1. (**a**) Construction of expression vector of *Deinococcus geothermalis* amylosucrase (DGAS). (**b**) SDS-PAGE analysis of recombinant DGAS expressed in *Escherichia coli* and purified on a nickel–nitrilotriacetic acid affinity column. (Lane M, protein molecular standard marker; Lane 1, crude enzyme; Lane 2, crude passing through enzyme; Lane 3, inclusion body; Lane 4, purified enzyme).

2.2. Transglycosylation of IFAs with DGAS

Three IFAs (daidzein, glycitein, and genistein) are found in soybeans, and their glucosides are linked by β-glycosidic bonds at C-7 (Figure 2) [4]. Glycosylation carried out by glycosyltransferases, such as AS, generates two types of *O*-glucosides, which are commonly defined as α- and β-linked glucosides [35]. The detailed glycosylation mechanism has not been clearly elucidated, but the glycosylation reaction usually occurs on the anomeric carbon with the attack of weak nucleophiles, such as the –OH group of sugars [35]. Glycosylation reactions often involve a unimolecular (S_N1) or bimolecular (S_N2) nucleophilic substitution at the anomeric center [35].

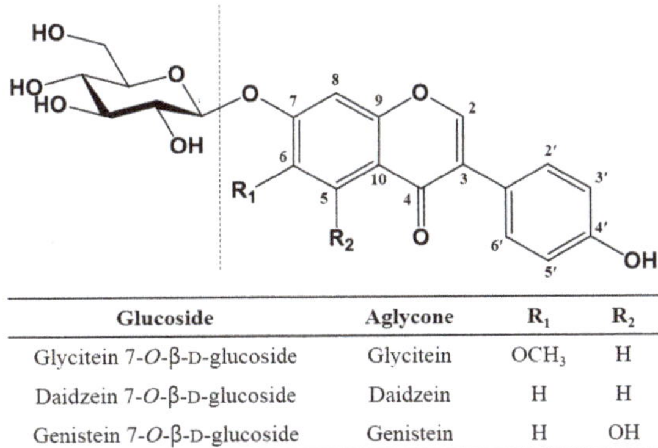

Glucoside	Aglycone	R_1	R_2
Glycitein 7-*O*-β-D-glucoside	Glycitein	OCH_3	H
Daidzein 7-*O*-β-D-glucoside	Daidzein	H	H
Genistein 7-*O*-β-D-glucoside	Genistein	H	OH

Figure 2. Structures of isoflavone glucosides (daidzin, glycitin, and genistin) present in soybeans.

DGAS used in this study produced various transglycosylated isoflavones with α-1,4-glycosidic linkages from IFAs (Figure 3). The transglycosylation of IFAs (such as in genistein) using DGAS are likely to occur at the C-7 or C-4′ positions rather than the C-5 position. Previous studies reported that recombinant AS created various stereostructures of acceptor compounds due to position-specific transglycosylation [27,33] (Figure 3). Moreover, the transglycosylation of flavonoids using DGAS has been reported to occur primarily at the C-7 position of the A ring and at the C-4′ position of the B ring [27,32,36]. DGAS has been reported to attach four or fewer glucose moieties to flavonoids in transglycosylation [21,32]. Thus, we expected the total number of glucose moieties attached to IFAs in the DGAS transglycosylation process to be four. For example, if four glucoses bind to the C-7 position, the number of glucoses that can attach to the C-4′ position is zero, and if three glucoses bind to the C-4′ position, one additional glucose may bind to the C-7 position (Figure 3).

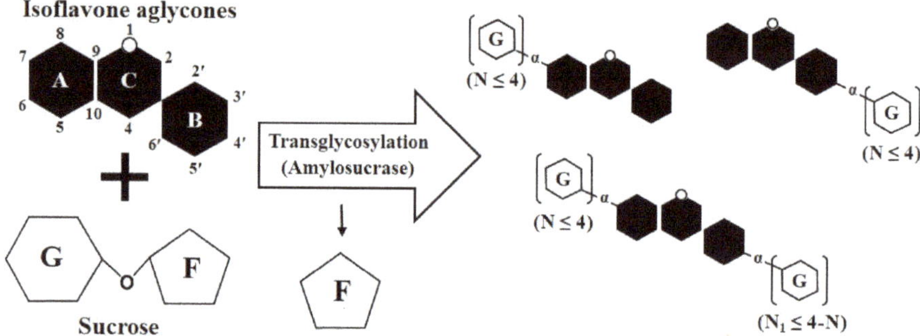

Figure 3. Scheme of transglycosylation reactions of isoflavone aglycones (IFAs) and sucrose using DGAS. G, glucose; F, fructose.

To identify reactants and products after transglycosylation, 12 isoflavone standards found in soybeans were separated by high-performance liquid chromatography (HPLC; peaks labeled: GD, daidzin; GL, glycitin; GG, genistin; MD, malonyldaidzin; ML, malonylglycitin; AD, acetyldaidzin; AL, acetylglycitin; MG, malonylgenistin; DN, daidzein; LN, glycitein; AG, acetylgenistin; GN, genistein) (Figure 4a). The results of transglycosylation using DGAS of the three IFA standards are shown chromatographically in Figure 4b–d. Compared to the 12 isoflavone standards, six to eight new peaks were detected, indicating that various transglycosylated products were produced from each aglycone during the enzymatic process (Figure 4b–d). Transglycosylated products from daidzein tended to be more polar (faster retention time) than transglycosylated glycitein and genistein glucosides for the same HPLC analysis conditions. This tendency was the same as the elution order of the three aglycones: daidzein > glycitein > genistein (Figure 4a). We attributed the different elution order of transglycosylated IFAs to their polarity differences, which are mainly influenced by the polarity of the parent compounds (acceptors in this study). If patterns of bound positions and the number of glucose moieties were the same in each aglycone, the elution order of transglycosylated IFAs according to increasing retention time would have been as follows: transglycosylated daidzein > transglycosylated glycitein > transglycosylated genistein. It was previously reported that α-glycosylisoquercitrin with more glucose moieties produced by the enzymatic transglycosylation of isoquercitrin eluted earlier than α-glycosylisoquercitrin with fewer glucose moieties [37]. Furthermore, after the transglycosylation of daidzin using DGAS, daidzein triglucoside eluted faster than daidzein diglucoside [36]. Therefore, in this study, transglycosylated IFAs with more glucose moieties had a faster elution time in reversed-phase HPLC analysis than transglycosylated IFAs with fewer glucose moieties (Figure 4b–d).

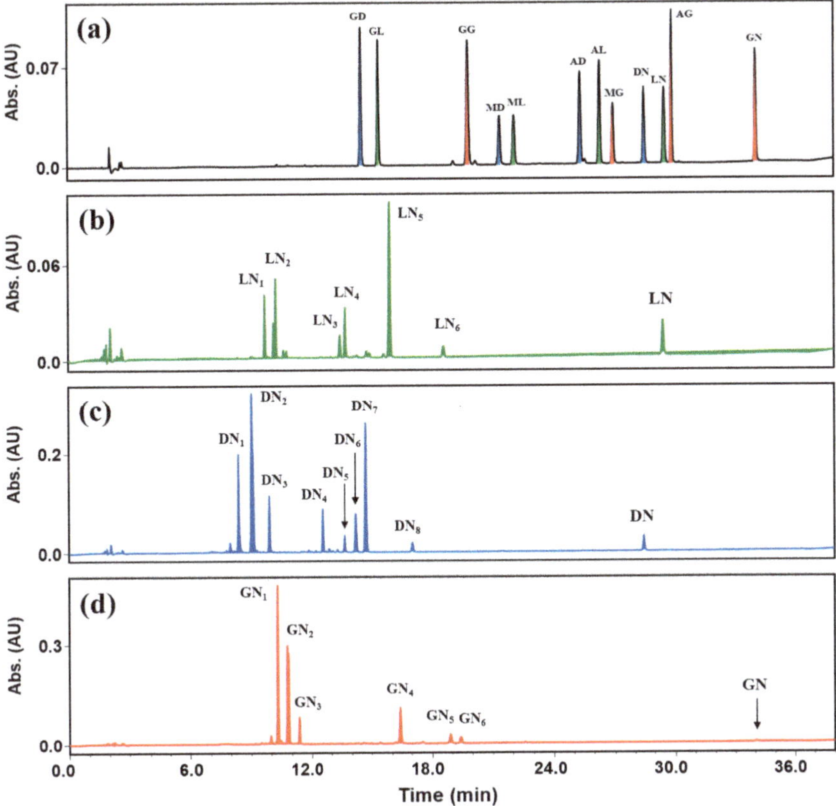

Figure 4. High-performance liquid chromatography (HPLC) traces (254 nm) of IFA standards after transglycosylation using DGAS. (**a**) Isoflavone standard mixture, (**b**) transglycosylated glycitein, (c) transglycosylated daidzein, and (**d**) transglycosylated genistein. Peaks: GD, daidzin; GL, glycitin; GG, genistin; MD, malonyldaidzin; ML, malonylglycitin; AD, acetyldaidzin; AL, acetylglycitin; MG, malonylgenistin; DN, daidzein; LN, glycitein; AG, acetylgenistin; GN, genistein. Peaks labeled LN_1 to LN_6, DN_1 to DN_8, and GN_1 to GN_6 are transglycosylated products from each IFA. Enzymatic transglycosylation reactions were performed using 1.0 mM of glycitein, 5.0 mM of daidzein, or 5.0 mM of genistein with 5.0 U of DGAS and 2.0 M of sucrose.

Peaks observed at the very beginning of elution (up to 12 min of retention time in this study) of the chromatograms were considered to be transglycosylated IFAs with two, three, or four glucose moieties (Figure 4b–d). However, more polar products (transglycosylated IFAs with 2–4 glucose moieties) among the transglycosylated IFAs resulted in lower resolution in the chromatogram. It was previously reported that transglycosylated products generated from the DGAS-catalyzed transglycosylation of isoquercitrin had similar retention times to each other, which resulted in very low resolution and greater difficultly separating them [32]. In addition, some isoquercitrin glucosides with high polarity have been reported to exist as isomers with the same number of sugars, the molecular weights of which were the same in mass spectra [32]. Based on previously reported results [32], we hypothesized that the transglycosylated IFA glucosides that eluted at similar retention times in this study were likely to be isomers with the same number of glucose moieties bound at different positions.

Peaks (except for minor peaks < 5% of the total product area) of transglycosylated reaction products from glycitein, daidzein, and genistein after enzymatic modification were labeled (Figure 4b–d). Peak LN_5 in Figure 4b had a retention time similar to glycitin (glycitein 7-β-*O*-glucoside; peak GL of Figure 4a).

The retention times of peak DN_7 in Figure 4c and peak GN_6 in Figure 4d were similar to those of daidzin (daidzein 7-β-*O*-glucoside; peak GD of Figure 4a) and genistin (genistein 7-β-*O*-glucoside; peak GG of Figure 4a), respectively. Thus, we assumed that peaks LN_5, DN_7, and GN_6 were α-glucosides of IFA with one glucose moiety attached. Standard compounds (daidzin, glycitin, and genistin) found in soybeans have a β-glycosidic bond between IFA and glucose. When IFAs are transglycosylated using DGAS, glucose moieties bound to the −OH group at the C-7 or C-4′ position in IFAs are present in an α-glycosidic bond with IFAs [27,33]. The 7-α-monoglucosides (peaks LN_5, DN_7, and GN_6) from the enzymatic modification of IFAs had different but similar retention times compared to standard isoflavone 7-β-monoglucosides (Figure 4b–d). Previously, daidzein triglucoside was reported to be more polar than daidzein diglucoside [36]. Due to their greater number of transglycosylated glucose moieties, LN_1–LN_4 from glycitein transglycosylation, DN_1–DN_6 from daidzein transglycosylation, and GN_1–GN_5 from genistein transglycosylation had higher polarities than LN_5, DN_7, and GN_6, respectively (Figure 4b–d). DGAS produces higher amounts of flavonoid glucosides attached with a greater number of glucose molecules to parent flavonoids such as isoquercitrin and daidzin than glycosyltransferases such as CGTase and amyloglucosidase [27,32,36], which produce higher levels of flavonoid glucosides transglycosylated with fewer sugar moieties [22,24]. When UDP-glucose is used as a donor for glycosidic binding, only one or two new transglycosylated products are produced by glycosyltransferases [16,22]. In contrast, DGAS generated a variety of new transglycosylated products from the IFAs (Figure 4b–d). As shown in Figure 3, the number of transglycosylated glucoses has been reported to range from one to four [27,32], because the free hydroxyl (−OH) groups of flavonoid aglycones and their transglycosylated glucosides can be used as potential transglycosylation sites in the DGAS enzyme process.

2.3. *Effects of Reaction Conditions on DGAS Transglycosylation*

Table 1 shows the conversion yields of IFAs to transglycosylated IFA glucosides according to the concentrations of donor and acceptors as well as the amount of DGAS. Glycitein reacted at lower concentrations than daidzein and genistein because of its lower solubility [38]. Regardless of IFA and sucrose concentrations, the conversion yield of each IFA used in this study increased as the amount of DGAS used increased, whereas the conversion yield of each IFA generally decreased with increasing IFA concentration. In transglycosylation reactions, a large amount of enzymes ensures not only a fast reaction rate, but also an increased reaction yield [33]. However, at a low donor concentration (0.1 M) and high acceptor concentration (20 mM), the conversion yields for daidzein and genistein were the lowest during transglycosylation for the highest amount of enzymes (5.0 U). The ratio of donor to acceptor suggests that the conversion yield is affected even if a sufficient amount of enzyme is supplied. Similar to the results of this study (Table 1), previous studies reported that the conversion efficiency increased when the concentration of the donor was higher than that of the acceptor [31,39,40]. The enzyme-modified glycosylation of lipophilic compounds is known to be complex because of the low solubility of lipophilic compounds in aqueous systems [41]. To increase the conversion yield despite low solubility, various solvents such as ionic liquids have been used in enzyme reactions for flavonoid glucoside synthesis [13,41]. In this study, we used dimethyl sulfoxide (DMSO) as the solvent to dissolve all IFAs (acceptors). However, cloudiness was observed during the enzyme reaction when the amount of acceptor was increased in an aqueous system containing a donor, acceptor, buffer, and enzyme (data not shown). Acceptor solubility was found to be the most important factor determining the lowest conversion yields at the highest acceptor conditions (glycitein, 4 mM; daidzein and genistein, 20 mM) (Table 1). Therefore, it is reasonable to assume that the reason glycitein had the lowest conversion yield was its relatively lower solubility than that of daidzein and genistein.

Table 1. Conversion yields of soy isoflavone aglycones after *Deinococcus geothermalis* amylosucrase (DGAS)-treated transglycosylation according to concentrations of donor and acceptors and amounts of the DGAS enzyme.

DGAS (Units)	Sucrose (M)	Glycitein (mM)			Daidzein (mM)			Genistein (mM)		
		0.05	1.0	4.0	0.2	5.0	20	0.2	5.0	20
0.5	0.1	0.0 ± 0.0^{c} [1]	1.9 ± 1.7^{c}	3.9 ± 0.2^{b}	0.0 ± 0.0^{f}	4.3 ± 0.6^{de}	2.4 ± 0.4^{ef}	10.9 ± 2.3^{c}	3.1 ± 1.3^{de}	2.1 ± 0.1^{e}
	1.0	0.0 ± 0.0^{c}	6.8 ± 0.8^{a}	4.2 ± 0.2^{b}	20.0 ± 2.5^{b}	8.3 ± 1.0^{cd}	2.7 ± 0.1^{ef}	59.8 ± 0.4^{b}	5.3 ± 0.3^{d}	0.8 ± 0.0^{e}
	2.0	0.0 ± 0.0^{c}	6.5 ± 0.8^{a}	4.9 ± 0.4^{ab}	28.8 ± 3.0^{a}	9.2 ± 1.0^{c}	1.8 ± 0.2^{ef}	68.6 ± 1.0^{a}	3.5 ± 0.1^{de}	1.2 ± 0.0^{e}
1.0	0.1	0.0 ± 0.0^{d}	7.0 ± 0.2^{c}	6.0 ± 0.3^{c}	22.5 ± 3.2^{d}	12.0 ± 0.9^{e}	4.4 ± 1.1^{f}	19.3 ± 1.9^{d}	7.1 ± 0.6^{e}	1.3 ± 0.0^{e}
	1.0	13.3 ± 2.5^{b}	19.0 ± 1.7^{a}	8.8 ± 0.4^{bc}	50.0 ± 2.4^{b}	32.5 ± 0.7^{c}	6.9 ± 0.1^{f}	53.1 ± 4.2^{b}	28.5 ± 0.2^{c}	4.8 ± 0.0^{e}
	2.0	8.6 ± 3.6^{bc}	20.8 ± 0.4^{a}	23.2 ± 3.7^{a}	55.5 ± 2.2^{a}	32.4 ± 0.8^{c}	4.8 ± 0.2^{f}	62.5 ± 4.7^{a}	24.7 ± 0.2^{cd}	2.9 ± 0.2^{e}
5.0	0.1	20.8 ± 6.4^{e}	35.0 ± 1.6^{d}	28.2 ± 0.8^{de}	72.0 ± 1.5^{b}	65.0 ± 0.8^{c}	23.3 ± 0.3^{f}	77.3 ± 2.7^{b}	74.8 ± 0.5^{bc}	23.4 ± 0.0^{e}
	1.0	16.3 ± 1.5^{e}	69.5 ± 9.2^{b}	56.8 ± 2.2^{c}	97.2 ± 2.8^{a}	95.0 ± 0.1^{a}	50.5 ± 0.0^{d}	92.2 ± 5.1^{a}	96.6 ± 1.0^{a}	66.0 ± 0.2^{d}
	2.0	18.0 ± 2.8^{e}	88.8 ± 3.8^{a}	52.1 ± 1.3^{c}	97.5 ± 0.7^{a}	96.9 ± 0.2^{a}	36.9 ± 0.1^{f}	92.1 ± 3.2^{a}	98.2 ± 0.1^{a}	70.1 ± 0.1^{cd}

[1] Data are expressed as means ± standard deviations ($n = 3$). Means with different superscripted letters indicate that for the same amount of DGAS, each isoflavone aglycone differed significantly by the Tukey–Kramer honestly significant difference test ($p < 0.05$).

The conversion yields of daidzein tended to be similar to those of genistein for the various reaction conditions used in this study. Glycitein had a lower conversion yield than genistein and daidzein (Table 1). Transglycosylation has been reported to occur at the hydroxyl group of the acceptor, resulting in the formation of an *O*-glycosidic bond between the acceptor and donor [35]. Likewise, a correlation between the number of hydroxyl groups and reactivity after the transglycosylation of hydroxyflavonoids with DGAS was observed [27]. The average conversion yields of daidzein and genistein were approximately 96.7% and 94.8%, respectively, at the highest amount of enzyme (5.0 U), lower acceptor concentrations (0.2–5.0 mM), and higher donor concentrations (1.0–2.0 M). In general, the highest conversion yield is achieved when sufficient enzymes and sucrose are provided for transglycosylation [33]. However, excess sucrose has been reported to reduce conversion yields [32]. Therefore, it is essential to have an adequate ratio of donor and acceptor as well as a sufficient amount of enzyme to attain higher conversion yields.

2.4. Application of DGAS to IFA-Rich Extract from Soybeans

Ultrasound-assisted extraction is known by its high efficiency, high productivity, lower consumption of solvents, and eco-friendly process [42]. In addition, in this study, to prevent protein interference after the extraction of soybeans with organic solvent, proteins were precipitated by dehydration and aggregation using ice-cold acetone [43]. After the successful conversion of IFAs into transglycosylated IFAs, optimal reaction conditions established for the DGAS process were used for the transglycosylation of IFA-rich soybean extract (SBE) as the acceptor. IFAs in SBE existed at lower concentrations than the other isoflavones such as isoflavone glucosides (Figure 5a and Table 2).

Table 2. Content (mg/g powder) of isoflavones in soybean extract (SBE), cellulase (CE)-treated SBE (CE-SBE; isoflavone aglycone-rich extract), and CE-SBE transglycosylated using DGAS (CE-SBE-DGAS).

Isoflavone	SBE	CE-SBE	CE-SBE-DGAS
Glycitein	0.0 ± 0.0 [b] [1]	3.98 ± 0.2 [a]	0.0 ± 0.0 [b]
Daidzein	0.79 ± 0.13 [b]	41.98 ± 1.26 [a]	0.0 ± 0.0 [c]
Genistein	1.0 ± 0.03 [b]	53.63 ± 4.35 [a]	0.0 ± 0.0 [c]

[1] Data are expressed as means ± standard deviations ($n = 3$). Means with different superscripted letters in the same row differ significantly by the Tukey-Kramer honestly significant difference test ($p < 0.05$).

Isoflavones in soybeans are mainly present in malonyl forms when extracted under normal conditions such as neutral solvent, ambient temperature, and short extraction time [44]. Similarly, malonyl forms were found to be the major isoflavones in SBE in this study (Figure 5a). Peaks labeled MD, ML, and MG are malonyldaidzin, malonylglycitin, and malonylgenistin, respectively (Figures 4a and 5a,b). Therefore, 7-β-*O*-glucoses (with acetyl and malonyl moieties) bound to IFA should be hydrolyzed using cellulase (CE) prior to DGAS transglycosylation. Extraction under harsh conditions (low pH and high temperature) and microbial fermentation has been used previously to obtain IFAs from soybeans [44]. However, acid-modified extraction is not suitable for enzymatic glycosylation reactions and is not environmentally friendly. Conjugated isoflavones (malonylisoflavones, acetylisoflavones, and non-acylated isoflavone glucosides) are converted into their non-conjugated counterparts using enzymes such as glycosylase [45]. In this study, we obtained IFAs by removing β-1,4-linked glucose moiety from conjugated isoflavones using CE. After CE processing, malonylisoflavones were degraded into their corresponding aglycones. As the concentrations of conjugated daidzeins (peaks GD, MD, and AD), conjugated glyciteins (peaks GL, ML, and AL), and conjugated genisteins (peaks GG and MG) in SBE decreased (Figure 5a), the concentrations of their corresponding IFAs increased after CE treatment (Figure 5b). CE-treated soybean extract (CE-SBE; IFA-rich extract) had the highest content of genistein (53.9% of total), daidzein (42.2% of total), and glycitein (4.0% of total) compared with SBE and CE-SBE transglycosylated using DGAS (CE-SBE-DGAS) (Table 2). Consistent with the results of this study, the content of glycitein derivatives in soybeans and their processed products has been reported to be less than 5% of the total isoflavone content [5].

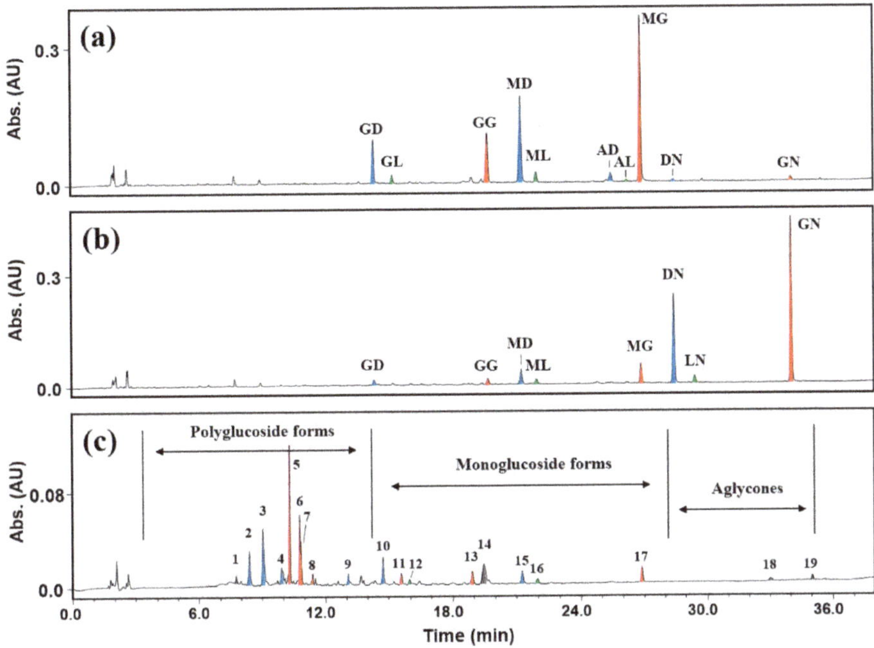

Figure 5. HPLC traces (254 nm) of IFA-rich soybean extract after transglycosylation using DGAS. (**a**) Soybean extract, (**b**) cellulase-treated soybean extract, (**c**) and cellulase-treated soybean extract transglycosylated using DGAS. Peaks: GD, daidzin; GL, glycitin; GG, genistin; MD, malonyldaidzin; ML, malonylglycitin; AD, acetyldaidzin; AL, acetylglycitin; MG, malonylgenistin; DN, daidzein; LN, glycitein; GN, genistein. Refer to Table 3 for the identification of each numbered peak of the 19 compounds.

Soy isoflavones have different ultraviolet-visible spectra depending on their aglycones [3,46]. The peaks newly produced by DGAS-treated transglycosylation were tentatively confirmed by comparison with the patterns of ultraviolet-visible spectra of isoflavone standards and molecular weight (Table 3). As results of mass spectrometry (MS), 19 products including four unknown compounds of DGAS-treated CE-SBE were tentatively identified (Table 3). The malonylisoflavones in SBE were not changed by DGAS (peaks 15, 16, and 17 of Figure 5c) in this study. In contrast, IFAs were not detected in CE-SBE-DGAS (Figure 5c and Table 2), indicating that IFAs in CE-SBE are successfully transglycosylated using DGAS. In previous studies, DGAS has been reported to transglycosylate a sugar moiety to the –OH group of the flavonoid aglycone [27,30,31]. It is also reported that glucose moiety was transferred to the isoflavone monoglucosides such as daidzin [36].

Among the peaks newly generated using DGAS, the peaks 10, 12, and 13 were identified as daidzein, glycitein, and genistein monoglucosides with different stereospecificity and retention time from daidzin, glycitin, and genistin, respectively (Figure 5c and Table 3). The generated isoflavone monoglucosides can be seen to have a similar polarity compared to the isoflavone 7-β-*O*-glucosides (peaks GD, GL, and GG) in the SBE (Figure 5a). The MS results of CE-SBE-DGAS indicate that the more polar components, which are eluted earlier than isoflavone monoglucosides, are transglycosylated with two or more glucose moieties (Figure 5c and Table 3). The treatment of IFA standards and IFA-rich extract with DGAS resulted in higher amounts of isoflavone polyglucosides than isoflavone monoglucosides (Figures 4b–d and 5c). As mentioned earlier, DGAS creates various transglycosylated isoflavones from IFAs. Studies of enzyme-modified flavonoids have reported binding of one or more glucose moieties to aglycones [24,27]. It has also been reported that glucose-added products, which are more polar than daidzin, are produced when transglycosylation is carried out using daidzin as the acceptor [36]. Hence, by using DGAS for the transglycosylation of CE-SBE, various isoflavone polyglucosides with two and three glucose moieties bound to their backbones, as well as isoflavone monoglucosides, can be generated.

Table 3. Identification of isoflavone derivatives of CE-SBE-DGAS.

Peak No.	Time (min)	λ_{max} (nm)	$[M]^+$ (*m/z*)	Aglycone	Identified Compound
1	7.74	219.6, 276.6	n.d [1]	Unknown	Unknown
2	8.38	248.1, 296.5	903.1	Daidzein	Daidzein tetraglucoside
3	9.04	248.1, 298.1	741.1	Daidzein	Daidzein triglucoside
4	9.91	248.1, 298.1	741.1	Daidzein	Daidzein triglucoside
5	10.30	257.1, 322.0	757.2	Genistein	Genistein triglucoside
6	10.77	257.1, 322.0	757.2	Genistein	Genistein triglucoside
7	10.83	257.6, 322.0	757.2	Genistein	Genistein triglucoside
8	11.38	257.6, 329.2	595.2	Genistein	Genistein diglucoside
9	13.08	248.1, 298.1	417.1	Daidzein	Daidzein diglucoside
10	14.77	245.7, 300.5	417.1	Daidzein	Daidzein monoglucoside
11	15.57	257.6, 322.0	595.2	Genistein	Genistein diglucoside
12	15.97	252.8, 319.6	447.1	Glycitein	Glycitein monoglucoside
13	18.90	257.6, 322.0	433.1	Genistein	Genistein monoglucoside
14	19.45	229.1, 271.8	n.d	Unknown	Unknown
15	21.25	248.1, 298.1	n.d	Daidzein	Malonyldaidzin
16	21.97	257.6, 317.2	n.d	Glycitein	Malonylglycitin
17	26.88	257.6, 324.4	n.d	Genistein	Malonylgenistin
18	33.02	267.1	n.d	Unknown	Unknown
19	35.04	233.8, 276.6	n.d	Unknown	Unknown

[1] n.d Out of the set molecular weight in mass conditions.

3. Materials and Methods

3.1. Chemicals

Sucrose, formic acid, DMSO, and sodium phosphate were purchased from Sigma-Aldrich Co., LLC (St. Louis, MO, USA). Daidzin, glycitin, genistin, daidzein, glycitein, and genistein were bought from Extrasynthese (Genay, France). Malonyldaidzin, malonylglycitin, malonylgenistin, acetyldaidzin, acetylglycitin, and acetylgenistin were purchased from Wako Chemical Co. (Tokyo, Japan). HPLC-grade water, acetonitrile, and methanol were obtained from Fisher Scientific (Hampton, NH, USA). All other chemicals used in this study were of analytical reagent grade.

3.2. Expression and Activity of DGAS

The *dgas* gene has been described in previous studies [21,30]. This gene was identified in the genome of *D. geothermalis* DSM 11300. *E. coli* MC1061 was used to express *dgas*, whereas *E. coli* DH10B was employed for gene manipulation. DGAS was expressed by *E. coli* MC1061 harboring *dgas* in the pHCXHD vector system, as described previously [32,36]. *Escherichia coli* transformed with pHCXHD-DGAS were incubated for 18 h and then harvested by centrifugation (Hanil Combi 514R; Hanil Centrifuge Co., Gimpo, Korea) at 7000× *g* for 20 min. Then, the supernatant was discarded, and the pellet was resuspended in lysis buffer (50 mM NaH_2PO_4, 300 mM NaCl, and 10 mM imidazole; pH 7.5) and disrupted using an iced ultrasonic bath (Sonifier 450; Branson Ultrasonics Corp., Danbury, CT, USA). Proteins extracted from the solution were collected by centrifugation (Hanil Combi 514R; Hanil Centrifuge Co.) at 7000× *g* at 4 °C for 20 min. Enzymes in the crude cell extract were purified with a nickel–nitrilotriacetic acid affinity column (Poly-Prep; Bio-Rad Laboratories, Inc., Hercules, CA, USA) filled with 500 μL nickel–nitrilotriacetic acid Superflow (Qiagen, Hilden, Germany). A purified enzyme was confirmed by performing sodium dodecyl sulfate–polyacrylamide gel electrophoresis using 10% (*v*/*v*) acrylamide. Protein concentrations were measured using a bicinchoninic acid assay kit (Thermo Fisher Scientific, Agawam, MA, USA) with bovine serum albumin as a standard.

The sucrose hydrolysis activity of DGAS was measured using 3,5-dinitrosalicylic acid solution, as described previously [32,36]. A reaction mixture comprising 50 μL 500 mM Tris-HCl (pH 8.0), 100 μL 25% (*w*/*v*) sucrose, and 300 μL deionized water was used. The enzymatic reaction was carried out by adding 50 μL of DGAS solution to the reaction mixture at 45 °C for 10 min. To stop the reaction, 500 μL of 3,5-dinitrosalicylic acid solution was added. After boiling for 5 min, the absorbance of the final reaction mixture was measured at 550 nm using a microplate reader (iMark™ Microplate Absorbance Reader; Bio-Rad Laboratories, Inc.). The reducing sugar concentration was calculated using a fructose standard curve. One unit of DGAS was defined as the amount of enzyme that produced one μmol of fructose per min under the assay conditions.

3.3. Transglycosylation of IFAs

Transglycosylation reactions with DGAS were performed using sucrose as the donor in sterilized water and IFA as an acceptor in DMSO. The reaction was performed in sodium phosphate buffer (pH 5.0 and 50 mM final concentration). Sucrose was used at concentrations of 10, 100, and 200 mM. Daidzein and genistein were used at concentrations of 20, 500, and 2000 μM, while glycitein was used at concentrations of 4, 100, and 400 μM. The amount of DGAS used was set to three enzyme units at 0.1, 1.0, and 5.0 U. The reaction temperature was 45 °C, and the reaction time was 24 h. These reaction conditions (temperature, buffer, and pH) for DGAS were based on those reported in previous studies [31,36]. The reaction was stopped by adding an equal amount of 10% (*v*/*v*) DMSO in methanol containing 0.1% (*v*/*v*) formic acid. The terminated reaction solution was subjected to vigorous mixing, sonicated for 10 min, and filtered through a 0.45 μm polyvinylidene fluoride syringe filter (GH Polypro; Pall Corp., Port Washington, NY, USA). Final samples were stored in the freezer at −20 °C prior to analysis.

3.4. Preparation and Transglycosylation of SBE

Soybean seeds (cv. Daewon) were obtained from the Department of Southern Area Crop Science, National Institute of Crop Science, Rural Development Administration (Dalseong, Korea). Soybean seeds (10 g) were pulverized using a cutting mill (Tube Mill 100 control; IKA®, Staufen, Germany). Extraction was carried out with 250 mL of 70% (*v*/*v*) aqueous ethanol using an ultrasonic bath (Sonifier 450; Branson Ultrasonics Corp.) at ambient temperature for 1 h. The aqueous ethanol extract of soybean seeds was filtered through Whatman no. 4 filter paper (Whatman Inc., Clifton, NJ, USA). The filtrate was evaporated using a rotary evaporator (N-1000; Eyela, Tokyo, Japan) in a water bath at 40 °C. Five hundred milliliters acetone was added to dry SBE, and this mixture was placed at −20 °C for 24 h. The mixture was filtered through Whatman no. 4 filter paper (Whatman Inc.) and evaporated (N-1000; Eyela). To obtain IFA-rich extract by enzymatically hydrolyzing conjugated isoflavones in SBE, 200 mg of SBE powder and 20 mg of CE (EC 3.2.1.6) (Vision Corp., Seoul, Korea) were reacted in water at 40 °C for 24 h. The reactant from the CE-SBE was concentrated in the methanol phase by removing the matrix using an ODS Sep-Pak cartridge (Oasis HLB; Waters, Milford, MA, USA). Purified CE-SBE reactant was evaporated using a rotary evaporator (N-1000; Eyela) in a water bath at 40 °C. For transglycosylation, CE-SBE powder (10 g/L) was used as an acceptor, and sucrose (100 mM) was used as the donor. The reaction temperature, time, buffer, and pH conditions of SBE were the same as those described in Section 3.3. After termination of the enzyme reaction, the pretreatment process for HPLC analysis proceeded in the same manner described in Section 3.3. The final samples were stored in a freezer at −20 °C until further analysis.

3.5. Analysis of Transglycosylated Isoflavones by HPLC and MS

SBE, CE-SBE, and CE-SBE-DGAS were analyzed using a Waters HPLC system (Alliance e2998; Waters) equipped with a ProntoSIL ace-EPS-C_{18} column (120 Å, 5 μm, 4.6 × 250 mm; Bischoff, Leonberg,

Germany) and a photodiode array detector (2695; Waters) at 254 nm. Gradient elution was performing using 0.1% (*v*/*v*) formic acid in water (solvent A) and acetonitrile (solvent B). All solvents were filtered, degassed, and kept under pressure. The initial mobile phase was 100% solvent A. The gradient of mobile phase B was as follows: 0–2.5 min, 0% B; 2.5–5 min, 0–12% B; 5–7 min, 12–18% B; 7–10 min, 18–18% B; 10–13 min, 18–26% B; 13–19 min, 26–26% B; 19–23 min, 26–46% B; 23–26 min, 46–46% B; 26–30 min, 46–75% B; 30–32 min, 75–0% B; 32–35 min, 0% B. Flow rate, column oven temperature, and injection volume were 1.0 mL/min, 30 °C, and 5 μL, respectively.

The mass detection was measured using the modified method used in our previous study [36]. The quadrupole Dalton-based (QDa) detector (Waters) was used to obtain the MS data. An isocratic solvent manager system split the analyte phase into acetonitrile at a ratio of 8:2. The QDa parameters in positive ion mode were as follows: capillary voltage, 0.8 kV; cone voltage, 5 V; source temperature, 600 °C; desolvation gas flow, 800 L/h. The mass values to identify the transglycosylated isoflavones were used for daidzein glucosides (255.1, 417.1, 579.1, 741.1, and 903.1 *m*/*z*), glycitein glucosides (285.1, 447.1, 609.1, 771.1, and 933.1 *m*/*z*), and genistein glucosides (271.2, 433.2, 595.2, 757.2, and 919.2 *m*/*z*), respectively. Empower 3 (Waters) was used to control the HPLC-QDa system and analyze the data obtained.

3.6. Statistical Analysis

Data are displayed as means ± standard deviations of three replicate determinations. Statistical analyses were performed using IBM SPSS software version 23.0 (IBM SPSS Statistics Inc., Chicago, IL, USA). One-way analysis of variance was performed to assess the significance of differences in mean values. Significant differences were verified by the Tukey–Kramer honestly significant difference test ($p < 0.05$).

4. Conclusions

Using the DGAS enzymatic process, IFAs were modified to generate various forms of isoflavone glucosides with different polarities. In addition, transglycosylated IFAs had α-glycosidic bonds in the isoflavone backbone because of the enzymatic properties of DGAS. Different conversion yields of transglycosylation products were achieved for different reaction conditions. The most efficient transglycosylation resulted from a low acceptor concentration (IFAs), donor-high concentration (sucrose), and high DGAS enzyme activity. Applying the reaction conditions used in this study, IFA-rich extracts from SBE were obtained by an environmentally friendly process using CE. The treatment of IFA standards and IFA-rich extract with DGAS produced higher amounts of isoflavone polyglucosides than isoflavone monoglucosides. Taken together, the results of this study suggest that DGAS-treated transglycosylation changes physicochemical properties such as the bioactivity, solubility and stability of isoflavones in soybeans and produces soybean-based functional ingredients rich in transglycosylated IFA.

Author Contributions: Conceptualization, Y.S.J. and D.-H.S.; methodology, Y.-J.K., D.J., and M.-S.K.; validation, Y.S.J.; formal analysis, T.G.N., and C.-S.R.; data curation, C.-S.R.; investigation, D.J., M.-S.K., D.-H.S., and T.G.N.; writing—original draft preparation, Y.S.J., and Y.-J.K.; writing—review and editing, A.T.K., C.-S.P., and D.-O.K.; supervision, D.-O.K.; project administration, D.-O.K. All authors have read and agreed to the published version of the manuscript.

Funding: This research was funded by the Rural Development Administration Agenda of Korea, grant number "PJ01251204".

Conflicts of Interest: The authors declare no conflict of interest.

Abbreviations

AS	amylosucrase
CE	cellulase
CE-SBE	cellulase-treated soybean extract
CE-SBE-DGAS	CE-SBE transglycosylated using DGAS
CGTase	cyclodextrin glycosyltransferase
DGAS	*Deinococcus geothermalis* amylosucrase
DMSO	dimethyl sulfoxide
HPLC	high-performance liquid chromatography
IFA	isoflavone aglycone
MS	mass spectrometry
SBE	soybean extract
QDa	quadrupole Dalton-based
UDP-glucose	uridine diphosphate glucose

References

1. Becker-Reshef, I.; Barkera, B.; Humber, M.; Puricelli, E.; Sanchez, A.; Sahajpal, R.; McGaughey, K.; Justice, C.; Baruth, B.; Wu, B.; et al. The GEOGLAM crop monitor for AMIS: Assessing crop conditions in the context of global markets. *Glob. Food. Secur.* **2019**, *23*, 173–181. [CrossRef]
2. Veitch, N.C. Isoflavonoids of the leguminosae. *Nat. Prod. Rep.* **2013**, *30*, 988–1027. [CrossRef]
3. Ko, K.-P. Isoflavones: Chemistry, analysis, functions and effects on health and cancer. *Asian Pac. J. Cancer Prev.* **2014**, *15*, 7001–7010. [CrossRef]
4. Jackson, C.-J.C.; Dini, J.P.; Lavandier, C.; Rupasinghe, H.P.V.; Faulkner, H.; Poysa, V.; Buzzell, D.; DeGrandis, S. Effects of processing on the content and composition of isoflavones during manufacturing of soy beverage and tofu. *Process Biochem.* **2002**, *37*, 1117–1123. [CrossRef]
5. Zaheer, K.; Akhtar, M.H. An updated review of dietary isoflavones: Nutrition, processing, bioavailability and impacts on human health. *Crit. Rev. Food Sci. Nutr.* **2017**, *57*, 1280–1293. [CrossRef]
6. Muthyala, R.S.; Ju, Y.H.; Sheng, S.; Williams, L.D.; Doerge, D.R.; Katzenellenbogen, B.S.; Helferich, W.G.; Katzenellenbogen, J.A. Equol, a natural estrogenic metabolite from soy isoflavones: Convenient preparation and resolution of *R*- and *S*-equols and their differing binding and biological activity through estrogen receptors alpha and beta. *Bioorg. Med. Chem.* **2004**, *12*, 1559–1567. [CrossRef] [PubMed]
7. Stancanelli, R.; Mazzaglia, A.; Tommasini, S.; Calabrò, M.L.; Villari, V.; Guardo, M.; Ficarra, P.; Ficarra, R. The enhancement of isoflavones water solubility by complexation with modified cyclodextrins: A spectroscopic investigation with implications in the pharmaceutical analysis. *J. Pharm. Biomed. Anal.* **2007**, *44*, 980–984. [CrossRef] [PubMed]
8. Noguchi, A.; Saito, A.; Homma, Y.; Nakao, M.; Sasaki, N.; Nishino, T.; Takahashi, S.; Nakayama, T. A UDP-glucose: Isoflavone 7-*O*-glucosyltransferase from the roots of soybean (*Glycine max*) seedlings: Purification, gene cloning, phylogenetics, and an implication for an alternative strategy of enzyme catalysis. *J. Biol. Chem.* **2007**, *282*, 23581–23590. [CrossRef] [PubMed]
9. Szeja, W.; Grynkiewicz, G.; Rusin, A. Isoflavones, their glycosides and glycoconjugates. Synthesis and biological activity. *Curr. Org. Chem.* **2017**, *21*, 218–235. [CrossRef]
10. Moradi, S.V.; Hussein, W.M.; Varamini, P.; Simerska, P.; Toth, I. Glycosylation, an effective synthetic strategy to improve the bioavailability of therapeutic peptides. *Chem. Sci.* **2016**, *7*, 2492–2500. [CrossRef]
11. Nidetzky, B.; Gutmann, A.; Zhong, C. Leloir glycosyltransferases as biocatalysts for chemical production. *Acs Catal.* **2018**, *8*, 6283–6300. [CrossRef]
12. Yang, M.; Davies, G.J.; Davis, B.G. A glycosynthase catalyst for the synthesis of flavonoid glycosides. *Angew. Chem. Int. Ed.* **2007**, *46*, 3885–3888. [CrossRef] [PubMed]
13. De Winter, K.; Verlinden, K.; Křen, V.; Weignerová, L.; Soetaert, W.; Desmet, T. Ionic liquids as cosolvents for glycosylation by sucrose phosphorylase: Balancing acceptor solubility and enzyme stability. *Green Chem.* **2013**, *15*, 1949–1955. [CrossRef]
14. De Roode, B.M.; Franssen, M.C.R.; van der Padt, A.; Boom, R.M. Perspectives for the industrial enzymatic production of glycosides. *Biotechnol. Progr.* **2003**, *19*, 1391–1402. [CrossRef]

15. Wang, Y.; Xu, W.; Bai, Y.; Zhang, T.; Jiang, B.; Mu, W. Identification of an α-(1,4)-glucan-synthesizing amylosucrase from *Cellulomonas carboniz* T26. *J. Agric. Food Chem.* **2017**, *65*, 2110–2119. [CrossRef]
16. Wang, Z.; Wang, S.; Xu, Z.; Li, M.; Chen, K.; Zhang, Y.; Hu, Z.; Zhang, M.; Zhang, Z.; Qiao, X.; et al. Highly promiscuous flavonoid 3-*O*-glycosyltransferase from *Scutellaria baicalensis*. *Org. Lett.* **2019**, *21*, 2241–2545. [CrossRef]
17. Li, D.; Roh, S.-A.; Shim, J.-H.; Mikami, B.; Baik, M.-Y.; Park, C.-S.; Park, K.-H. Glycosylation of genistin into soluble inclusion complex form of cyclic glucans by enzymatic modification. *J. Agric. Food Chem.* **2005**, *53*, 6516–6524. [CrossRef]
18. Kometani, T.; Nishimura, T.; Nakae, T.; Takii, H.; Okada, S. Synthesis of neohesperidin glycosides and naringin glycosides by cyclodextrin glucanotransferase from an alkalophilic *Bacillus* species. *Biosci. Biotechnol. Biochem.* **1996**, *60*, 645–649. [CrossRef]
19. Sato, T.; Nakagawa, H.; Kurosu, J.; Yoshida, K.; Tsugane, T.; Shimura, S.; Kirimura, K.; Kino, K.; Usami, S. α-Anomer-selective glucosylation of (+)-catechin by the crude enzyme, showing glucosyl transfer activity, of *Xanthomonas campestris* WU-9701. *J. Biosci. Bioeng.* **2000**, *90*, 625–630. [CrossRef]
20. Křen, V.; Martínková, L. Glycosides in medicine: "The role of glycosidic residue in biological activity". *Curr. Med. Chem.* **2001**, *8*, 1303–1328. [CrossRef]
21. Cho, H.-K.; Kim, H.-H.; Seo, D.-H.; Jung, J.-H.; Park, J.-H.; Baek, N.-I.; Kim, M.-J.; Yoo, S.-H.; Cha, J.; Kim, Y.-R.; et al. Biosynthesis of (+)-catechin glycosides using recombinant amylosucrase from *Deinococcus geothermalis* DSM 11300. *Enzym. Microb. Technol.* **2011**, *49*, 246–253. [CrossRef] [PubMed]
22. Gantt, R.W.; Peltier-Pain, P.; Cournoyer, W.J.; Thorson, J.S. Using simple donors to drive the equilibria of glycosyltransferase-catalyzed reactions. *Nat. Chem. Biol.* **2011**, *7*, 685–691. [CrossRef] [PubMed]
23. Xiao, J.; Muzashvili, T.S.; Georgiev, M.I. Advances in the biotechnological glycosylation of valuable flavonoids. *Biotechnol. Adv.* **2014**, *32*, 1145–1156. [CrossRef] [PubMed]
24. Lee, Y.-S.; Woo, J.-B.; Ryu, S.-I.; Moon, S.-K.; Han, N.S.; Lee, S.-B. Glucosylation of flavonol and flavanones by *Bacillus* cyclodextrin glucosyltransferase to enhance their solubility and stability. *Food Chem.* **2017**, *229*, 75–83. [CrossRef] [PubMed]
25. Park, H.; Kim, J.; Park, J.-H.; Baek, N.-I.; Park, C.-S.; Lee, H.-S.; Cha, J. Bioconversion of piceid to piceid glucoside using amylosucrase from *Alteromonas macleodii* deep ecotype. *J. Microbiol. Biotech.* **2012**, *22*, 1698–1704. [CrossRef] [PubMed]
26. Del Valle, E.M.M. Cyclodextrins and their uses: A review. *Process Biochem.* **2004**, *39*, 1033–1046. [CrossRef]
27. Rha, C.-S.; Jung, Y.S.; Seo, D.-H.; Kim, D.-O.; Park, C.-S. Site-specific α-glycosylation of hydroxyflavones and hydroxyflavanones by amylosucrase from *Deinococcus geothermalis*. *Enzym. Microb. Technol.* **2019**, *129*. [CrossRef]
28. Skov, L.K.; Mirza, O.; Henriksen, A.; De Montalk, G.P.; Remaud-Simeon, M.; Sarçabal, P.; Willemot, R.-M.; Monsan, P.; Gajhede, M. Amylosucrase, a glucan-synthesizing enzyme from the α-amylase Family. *J. Biol. Chem.* **2001**, *276*, 25273–25278. [CrossRef]
29. Park, C.-S.; Park, I. The structural characteristics of amylosucrase-treated waxy corn starch and relationship between its in vitro digestibility. *Food Sci. Biotechnol.* **2017**, *26*, 381–387. [CrossRef]
30. Seo, D.-H.; Jung, J.-H.; Jung, D.-H.; Park, S.; Yoo, S.-H.; Kim, Y.-R.; Park, C.-S. An unusual chimeric amylosucrase generated by domain-swapping mutagenesis. *Enzym. Microb. Technol.* **2016**, *86*, 7–16. [CrossRef]
31. Kim, M.-D.; Jung, D.-H.; Seo, D.-H.; Jung, J.-H.; Seo, E.-J.; Baek, N.-I.; Yoo, S.-H.; Park, C.-S. Acceptor specificity of amylosucrase from *Deinococcus radiopugnans* and its application for synthesis of rutin derivatives. *J. Microbiol. Biotech.* **2016**, *26*, 1845–1854. [CrossRef] [PubMed]
32. Rha, C.-S.; Choi, J.-M.; Jung, Y.S.; Kim, E.-R.; Ko, M.J.; Seo, D.-H.; Kim, D.-O.; Park, C.-S. High-efficiency enzymatic production of α-isoquercitrin glucosides by amylosucrase from *Deinococcus geothermalis*. *Enzym. Microb. Technol.* **2019**, *120*, 84–90. [CrossRef] [PubMed]
33. Jang, S.-W.; Cho, C.H.; Jung, Y.-S.; Rha, C.; Nam, T.-G.; Kim, D.-O.; Lee, Y.-G.; Baek, N.-I.; Park, C.-S.; Lee, B.-H.; et al. Enzymatic synthesis of α-flavone glucoside via regioselective transglucosylation by amylosucrase from *Deinococcus geothermalis*. *PLoS ONE* **2018**, *13*, e0207466. [CrossRef] [PubMed]
34. Rizzo, G.; Baroni, L. Soy, soy foods and their role in vegetarian diets. *Nutrients* **2018**, *10*, 43. [CrossRef]
35. Demchenko, A.V. 1,2-*cis* *O*-Glycosylation: Methods, strategies, principles. *Curr. Org. Chem.* **2003**, *7*, 35–79. [CrossRef]

36. Kim, E.-R.; Rha, C.-S.; Jung, Y.S.; Choi, J.-M.; Kim, G.-T.; Jung, D.-H.; Kim, T.-J.; Seo, D.-H.; Kim, D.-O.; Park, C.-S. Enzymatic modification of daidzin using heterologously expressed amylosucrase in *Bacillus subtilis*. *Food Sci. Biotechnol.* **2019**, *28*, 165–174. [CrossRef]
37. Nyska, A.; Hayashi, S.-m.; Koyanagi, M.; Davis, J.P.; Jokinen, M.P.; Ramot, Y.; Maronpot, R.R. Ninety-day toxicity and single-dose toxicokinetics study of *alpha*-glycosyl isoquercitrin in Sprague-Dawley rats. *Food Chem. Toxicol.* **2016**, *97*, 354–366. [CrossRef]
38. Yatsu, F.K.J.; Koester, L.S.; Lula, I.; Passos, J.J.; Sinisterra, R.; Bassani, V.L. Multiple complexation of cyclodextrin with soy isoflavones present in an enriched fraction. *Carbohydr. Polym.* **2013**, *98*, 726–735. [CrossRef]
39. Overwin, H.; Wray, V.; Hofer, B. Biotransformation of phloretin by amylosucrase yields three novel dihydrochalcone glucosides. *J. Biotechnol.* **2015**, *211*, 103–106. [CrossRef]
40. Te Poele, E.M.; Valk, V.; Devlamynck, T.; van Leeuwen, S.S.; Dijkhuizen, L. Catechol glucosides act as donor/acceptor substrates of glucansucrase enzymes of *Lactobacillus reuteri*. *Appl. Microbiol. Biotechnol.* **2017**, *101*, 4495–4505. [CrossRef]
41. Bertrand, A.; Morel, S.; Lefoulon, F.; Rolland, Y.; Monsan, P.; Remaud-Simeon, M. *Leuconostoc mesenteroides* glucansucrase synthesis of flavonoid glucosides by acceptor reactions in aqueous-organic solvents. *Carbohydr. Res.* **2006**, *341*, 855–863. [CrossRef] [PubMed]
42. Galviz-Quezada, A.; Ochoa-Aristizábal, A.M.; Zabala, M.E.A.; Ochoa, S.; Osorio-Tobón, J.F. Valorization of iraca (*Carludovica palmata*, Ruiz & Pav.) infructescence by ultrasound-assisted extraction: An economic evaluation. *Food. Bioprod. Process.* **2019**, *118*, 91–102.
43. Fic, E.; Kędracka-Krok, S.; Jankowska, U.; Pirog, A.; Dziedzicka-Wasylewska, M. Comparison of protein precipitation methods for various rat brain structures prior to proteomic analysis. *Electrophoresis* **2010**, *31*, 3573–3579. [CrossRef] [PubMed]
44. Lin, F.; Giusti, M.M. Effect of solvent polarity and acidity on the extraction efficiency of isoflavones from soybeans (*Glycine max*). *J. Agric. Food Chem.* **2005**, *53*, 3795–3800. [CrossRef] [PubMed]
45. Simmons, A.L.; Chitchumroonchokchai, C.; Vodovotz, Y.; Failla, M.L. Isoflavone retention during processing, bioaccessibility, and transport by Caco-2 cells: Effects of source and amount of fat in a soy soft pretzel. *J. Agric. Food Chem.* **2012**, *60*, 12196–12203. [CrossRef] [PubMed]
46. Sun, J.-M.; Sun, B.-L.; Han, F.-X.; Yan, S.-R.; Yang, H.; Akio, K. Rapid HPLC method for determination of 12 isoflavone components in soybean seeds. *Agric. Sci. China* **2011**, *10*, 70–77. [CrossRef]

Sample Availability: Samples of the compounds are not available from the authors.

Article

Conversion of Oleic Acid into Azelaic and Pelargonic Acid by a Chemo-Enzymatic Route

Elisabetta Brenna [1,*], Danilo Colombo [1], Giuseppe Di Lecce [2], Francesco G. Gatti [1], Maria Chiara Ghezzi [1], Francesca Tentori [1], Davide Tessaro [1] and Mariacristina Viola [1,†]

1 Dipartimento di Chimica, Materiali ed Ingegneria Chimica "Giulio Natta", Politecnico di Milano, Via Mancinelli 7, 20131 Milano, Italy; danilo.colombo@polimi.it (D.C.); francesco.gatti@polimi.it (F.G.G.); mariachiara.ghezzi@polimi.it (M.C.G.); francesca.tentori@polimi.it (F.T.); davide.tessaro@polimi.it (D.T.); mariacristina.viola@polimi.it (M.V.)

2 Oleificio Zucchi S.p.A., Via Acquaviva, 26100 Cremona, Italy; dirlab@oleificiozucchi.com

* Correspondence: mariaelisabetta.brenna@polimi.it; Tel.: +39-02-2399-3077

† The authors are listed in alphabetical order.

Academic Editor: Josefina Aleu
Received: 30 March 2020; Accepted: 16 April 2020; Published: 18 April 2020

Abstract: A chemo-enzymatic approach for the conversion of oleic acid into azelaic and pelargonic acid is herein described. It represents a sustainable alternative to ozonolysis, currently employed at the industrial scale to perform the reaction. Azelaic acid is produced in high chemical purity in 44% isolation yield after three steps, avoiding column chromatography purifications. In the first step, the lipase-mediated generation of peroleic acid in the presence of 35% H_2O_2 is employed for the self-epoxidation of the unsaturated acid to the corresponding oxirane derivative. This intermediate is submitted to in situ acid-catalyzed opening, to afford 9,10-dihydroxystearic acid, which readily crystallizes from the reaction medium. The chemical oxidation of the diol derivative, using atmospheric oxygen as a stoichiometric oxidant with catalytic quantities of $Fe(NO_3)_3 \cdot 9 \cdot H_2O$, (2,2,6,6-tetramethylpiperidin-1-yl)oxyl (TEMPO), and NaCl, affords 9,10-dioxostearic acid which is cleaved by the action of 35% H_2O_2 in mild conditions, without requiring any catalyst, to give pelargonic and azelaic acid.

Keywords: lipase; biocatalysis; unsaturated fatty acid; oxidative cleavage; oxidation

1. Introduction

The employment of renewable feedstocks in the chemical industry is steadily advancing to ensure more efficient use of natural resources, reduce the dependence on fossil raw materials, and give a contribution to achieving sustainable consumption and production patterns [1].

Fats and oils represent an important class of renewable feedstock from which the so-called oleochemicals are obtained. They are abundant in nature, biodegradable, and have nontoxic properties. They have long hydrocarbon chains, resembling the structure of petroleum components, but they are also characterized by several functional groups useful for chemical modification. The major process for transforming fats and oils into oleochemicals is hydrolysis, converting natural triglycerides into crude glycerine and mixtures of fatty acids. The latter are then submitted to reactions involving either the carboxylic group (to afford soaps, esters, amides, amines, and alcohols) or the reduction/oxidation of the C=C double bonds, if present. Among these procedures developed to obtain fine chemicals, the oxidative cleavage of unsaturated fatty acids for the production of dicarboxylic acids, hydroxy acids, and amino acids has received great attention in the last decade [2–4]. Until recently, only two dicarboxylic acids prepared from oleochemicals have been commercialized, i.e., sebacic acid (**1**), obtained by the alkaline cleavage of castor oil [5], and azelaic acid (**2**), which is produced together with pelargonic acid (**3**) by ozonolysis of oleic acid (**4**) (Figure 1) [6]. Sebacic and azelaic acid are extensively employed in the synthesis of new generation biodegradable copolymers [7]. Azelaic acid, naturally occurring in wheat, rye, and barley, also finds application as an active ingredient in products for the topical treatment of acne [8], and for the stimulation of hair regrowth [9]. It works by inhibiting the growth of skin bacteria causing acne, and by keeping skin pores clear. Pelargonic acid, found in nature as ester derivatives in the oil of pelargonium, is used as an herbicide to prevent the growth of weeds both indoors and outdoors, and as a blossom thinner for apple and pear trees [10].

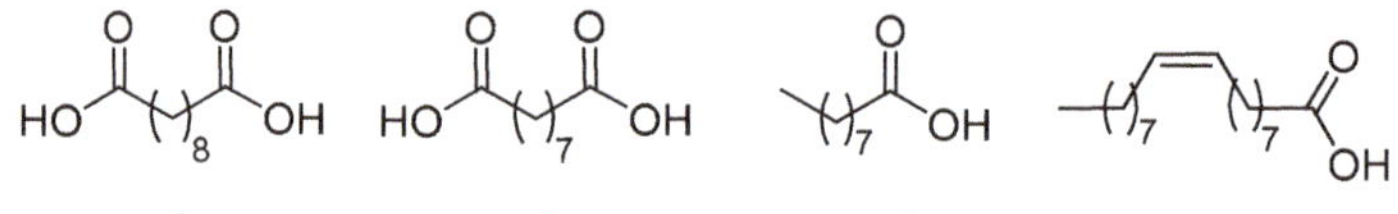

Figure 1. Sebacic acid (**1**), azelaic acid (**2**), pelargonic acid (**3**), and oleic acid (**4**).

Oleic acid is the most abundant monounsaturated fatty acid in nature [11], present in a wide range of vegetable and animal oils and fats. Several works have been published in recent years describing alternative methods to the ozone-promoted oxidative scission, most of which are based on metal catalysis [12]. Among them, some very effective one-pot procedures involve the use of H_2O_2 as primary oxidant in the presence of tungsten derivatives: (i) methyltrioctylammonium tetrakis(oxodiperoxotungsto)phosphate [13] (40% H_2O_2, 85 °C, yields of compounds **2** and **3** by GC/MS analysis of the crude mixture were 79% and 82%, respectively); (ii) WO_3 and Na_2SnO_3 in *t*-BuOH [14] (31% H_2O_2, 130 °C, sealed glass vial, isolation yields were 89% and 65% for **2** and **3**, respectively); (iii) a new hybrid organic/inorganic polyoxotungstate in *t*-BuOH [15] (30% H_2O_2, 120 °C, yield of compounds **2** and **3** by GC/MS analysis of the crude mixture were 79% and 80%, respectively); iv) H_2WO_4 [16] (60% H_2O_2, reflux, isolation yield was 60% for **2**); (v) Na_2WO_4 aqueous solution/H_3PO_4 aqueous solution with a suitable phase transfer catalyst in a sealed flask [17] (30% H_2O_2, 90 °C, yield of compound **2** by GC/MS analysis of the crude mixture was max 54%); (vi) the peroxo–tungsten complex $[C_5H_5N(n\text{-}C_{16}H_{33})]_3\{PO_4[WO(O_2)_2]_4\}$ as a phase-transfer catalyst/co-oxidant [18] (30% H_2O_2, 85 °C, yields of compounds **2** and **3** by GC/MS analysis of the crude mixture were 79% and 80%, respectively), using in this case oleic acid obtained upon hydrolysis of high oleic sunflower oil by *Candida cylindracea* lipase.

As for biocatalytic methods, only a few reports have appeared in the literature. Song et al. developed [19,20] a multi-step enzymatic procedure (Figure 2) starting from the hydration of oleic acid (**4**), followed by the oxidation of the intermediate alcohol **5** to the ketone derivative **6** that was in turn submitted to Baeyer–Villiger (BV) oxidation to afford ester **7**. The latter was hydrolyzed to afford acid **3** and the hydroxy acid **8**, which was finally oxidized to azelaic acid (**2**).

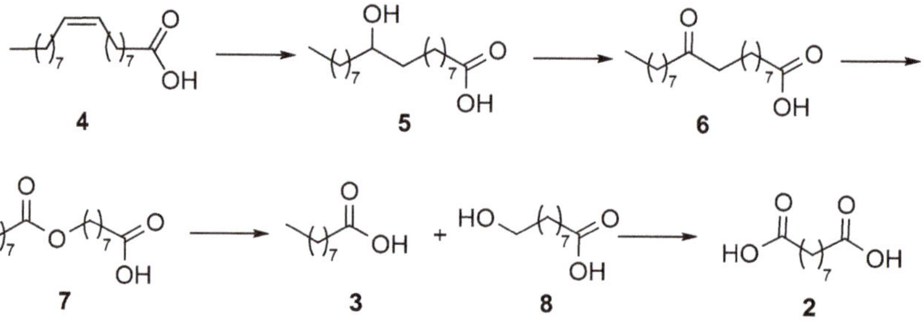

Figure 2. Synthesis of azelaic acid (**2**) from oleic acid (**4**) according to references [19,20].

In [19] the possibility to obtain a regioisomer of ester **7** giving directly diacid **2** upon hydrolysis was described, but it was not considered for further study and optimization in the following publication [20]. The same research group described the preparation of azelaic acid from ricinoleic acid [21]. The use of linoleic acid for the biocatalyzed production of azelaic acid was reported by Hauer et al. [22,23]. A multi-enzymatic one-pot reaction was developed to convert linoleic acid into azelaic acid by combining a 9S-lipoxygenase and 9/13-hydroperoxide lyase to obtain 9-oxononanoic acid submitted to the final oxidation to acid **2** catalyzed by an alcohol dehydrogenase. In 2019 the capability of *Candida tropicalis* ATCC20962 to transform nonanoic acid and its esters into azelaic acid **2** with the aid of nonane addition and continuous glucose supply [24] was investigated, to improve the production yield of diacid **2** obtained in the ozonolysis process of oleic acid.

Recently, we were involved in a project aimed at the valorization of the side-stream products generated by an Italian plant for vegetable seed oil refining (Oleificio Zucchi, Cremona) by using biocatalytic methods. An important step of the refining process (neutralization) is represented by the removal of free fatty acids, producing a side-product, called soapstock, which is currently disposed of by Zucchi in bio-digesters. The fatty acids profile of this material depends on the nature of the vegetable oil, and, in particular, the one obtained from sunflower oil is highly enriched in oleic acid [25]. Thus, we started to investigate a novel chemo-enzymatic oxidative scission of oleic acid, to be applied to the sunflower soapstock coming from Oleificio Zucchi for its valorization. The preliminary results of this study were obtained while working on commercial oleic acid, as a model compound, to study each step of the most suitable synthetic procedure more easily using a less complex starting material. The results are herein reported.

2. Results

The enzymatic synthesis of azelaic acid reported in 2013 by Song et al. [19] (Figure 2) consisted of the use of recombinant *Escherichia coli* cells expressing at the same time the genes encoding an oleate hydratase from *Stenotrophomonas maltophilia*, an alcohol dehydrogenase (ADH) from *Micrococcus luteus*, and a BV monooxygenase (BVMO) from *Pseudomonas putida* KT2440 for the transformation of oleic acid into 9-(nonanoyloxy)nonanoic acid (**7**). The hydrolysis of this latter compound, mediated by a cell extract of *E. coli* expressing the esterase gene from *P. fluorescens*, afforded pelargonic acid (**3**) and 9-hydroxynonanoic acid (**8**). In a further development of the work [20], the oxidation of derivative **8** by an ADH from *P. putida* GPo1 completed the route to azelaic acid. As the final product concentration in the reaction medium was only a few millimolar, likely because of the toxic effects of pelargonic acid on the *E. coli* cells, the authors investigated the hydrolysis of **7** and the subsequent oxidation of derivative **8** into acid **2** by chemical methods [26]. The ester intermediate **7** was purified by extraction and column chromatography, hydrolyzed with sodium hydroxide in methanol/water (4/1) at 60 °C to afford 9-hydroxynonanoic acid **8**, which was separated from pelargonic acid by column chromatography. Finally, the oxidation of the terminal hydroxy group of derivative **8** was achieved using $NaClO_2$ (1.2 equiv.), 2,2,6,6-tetramethyl-piperidin-1-yl oxyl (TEMPO) (4 mol%), and NaOCl (2 mol%) in aqueous acetonitrile. After these two steps, no purification was needed. The overall molar yield of azelaic acid from oleic acid was 58%.

We adopted a different strategy (Figure 3), consisting of the epoxidation of oleic acid to derivative **9**, followed by the formation of *threo*-9,10-dihydroxystearic acid (**10**) due to the acid-catalyzed hydrolytic cleavage of the oxirane ring, promoted directly in the epoxidation medium. The chemical oxidation of the diol afforded 9,10-dioxostearic acid (**11**), which was submitted to oxidative cleavage to afford a mixture of pelargonic (**3**) and azelaic (**2**) acid, through the intermediate anhydride **12**. In a recent publication [27], the oxidation of the methyl ester derivative of compound **10**, using a solvent-free procedure of dehydrogenative oxidation catalyzed by commercial 64 wt.% Ni/SiO_2 in the presence of 1-decene as a scavenger, was employed to afford a mixture of the two possible regioisomeric vicinal ketols, that were successively cleaved with formic acid/hydrogen peroxide, and afforded up to 80% pelargonic acid and azelaic acid monomethyl ester.

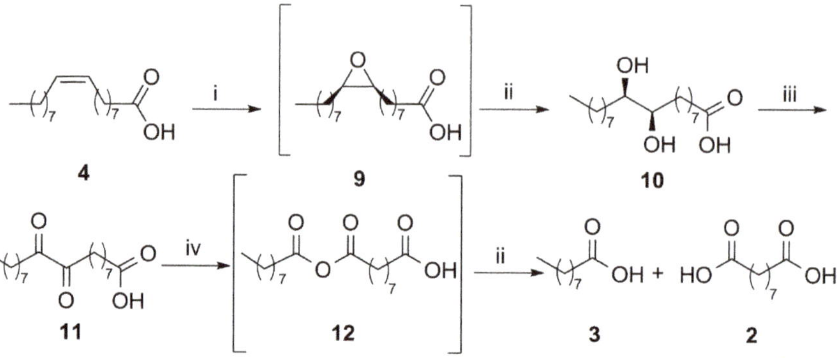

Figure 3. The synthesis of azelaic acid (**2**) from oleic acid (**4**) described in this paper. (i) H_2O_2 35%, Novozyme 435, acetonitrile, 5 h, 50 °C; (ii) $NaHSO_3$ saturated solution, H_2SO_4 2 M, 3 h, r.t.; (iii) atmospheric O_2, cat. $Fe(NO_3)_3 \cdot 9\ H_2O$/TEMPO/NaCl, toluene, 5 h, 100 °C; (iv) 35% H_2O_2, toluene, 3 h, 30 °C.

2.1. *Epoxidation of Oleic Acid* (**4**) *to 8-(3-Octyloxiran-2-yl)Octanoic Acid* (**9**)

The capability of certain lipases to catalyze the perhydrolysis (i.e., lysis by hydrogen peroxide) of carboxylic acid esters, hence forming peroxycarboxylic acids in aqueous hydrogen peroxide solutions, had been already patented by Clorox co. in the late eighties [28]. In 1990, an immobilized form of lipase B from *Candida antarctica* (Novozyme 435) was shown [29] to catalyze the formation of peroxycarboxylic acids directly from the corresponding carboxylic acid. In that case, the reaction was combined with the epoxidation of alkenes. A few years later, Warwel et al. [30] described that, when unsaturated fatty acids (or their esters) are treated with hydrogen peroxide in the presence of Novozyme 435, epoxidized derivatives are obtained through two sequential steps. Firstly, unsaturated fatty acids are converted into unsaturated peroxy acids by lipase-catalyzed perhydrolysis. Unsaturated peroxy or carboxylic acids are in turn epoxidized via a classical Prileshajev reaction, which is, in this case, referred to as a "self-epoxidation reaction" even though it proceeds predominantly via an intermolecular process.

The reaction has been widely exploited not only for the epoxidation of fatty acids and esters but also for the derivatization of vegetable oils [31]. Typically, the reaction medium consists of an aqueous layer containing hydrogen peroxide, an organic layer (usually a toluene solution) containing the fatty acid derivative, and a solid phase represented by the immobilized enzyme. The main issue of this chemo-enzymatic procedure is the deactivation of lipase. Temperature, reaction time, concentration of H_2O_2 in the reaction medium, and the related concentration of peracid generated in situ are critical parameters to be considered. Temperatures not higher than 50 °C, diluted H_2O_2 solution (max 1% *w/w* in the final solution) and reaction times not longer than 6 h represent the most common experimental conditions.

We decided to perform the chemo-enzymatic epoxidation of oleic acid in a water-miscible solvent, such as acetonitrile, for the following reasons: i) to promote the dissolution of both oleic acid and H_2O_2 in the same medium, and ii) to enable the in situ acid-catalyzed hydrolysis of the epoxide derivative at the end of the reaction, after removal of the enzyme, by addition of a diluted solution of sulfuric acid. The preliminary experiments were carried out with 0.30 mmol of commercial oleic acid (91% purity by GC/MS, the major contaminants are palmitic and stearic acid), changing the molar ratio H_2O_2/oleic acid (1.8 and 2.2), the temperature (30 °C and 50 °C), the amount of Novozyme 435 (10 mg and 30 mg), and the solvent volume (2 mL and 6 mL). The reactions were monitored by GC/MS. The results of this screening are reported in Tables S1 and S2 (see Supplementary Material). The following conditions were found to be optimal for running the reaction: 0.15 M oleic acid, 0.27 M H_2O_2, 5 mg·mL^{-1} Novozyme 435, in acetonitrile at 50 °C for 5 h under stirring with final 98% conversion (GC/MS,). Epoxide **9** could be recovered from the reaction mixture at 83% isolation yield (see Supplementary Material), starting from commercial oleic acid. The stereochemistry of the oxirane ring of derivative **9** was based on the *cis* configuration of oleic acid and was confirmed by comparison with literature data (see Supplementary Material).

2.2. *Acid–Catalyzed Cleavage of 8-(3-Octyloxiran-2-yl)Octanoic Acid* (**9**) *to 9,10-Didhydroxystearic Acid* (**10**)

The ring-opening of epoxystearic acid **9** to the corresponding *threo* diol derivative **10** (Figure 3) was promoted by diluted sulfuric acid, and the reaction conditions were adjusted to obtain quantitative conversion after 3 h at room temperature (Table S3 of Supplementary Material). The relative configuration of diol **10** was established following the *anti* mechanism of the hydrolytic opening of the oxirane ring of *cis*-epoxide **9** and was confirmed by literature data (see Materials and Methods). The acid-catalyzed hydrolysis was then performed, without isolation of the epoxide, by addition of a 2.0 M solution of H_2SO_4 directly to the reaction mixture of the epoxidation step, after removal of the enzyme by filtration and decomposition of peroxy species. This one-pot two-steps sequence led to the spontaneous crystallization of dihydroxystearic acid **10**, which was easily recovered from the reaction medium by filtration as a pure white solid. Satisfactory isolation yields (77%) were obtained, even when the reaction was run at a 2 g scale, starting from commercial 91% oleic acid.

2.3. *Oxidation of 9,10-Dihydroxystearic Acid* (**10**) *to 9,10-Dioxostearic Acid* (**11**)

The following procedures were tested to achieve the oxidation of either one or both the hydroxy groups of derivative **10**: (i) alcohol dehydrogenase—mediated oxidation (commercial kit from EVOXX); (ii) chemo-enzymatic oxidation with laccase and hydroxybenzotriazol (HOBt); (iii) aerobic oxidation with catalytic $Fe(NO_3)_3{\cdot}9\,H_2O$, TEMPO, and NaCl. Only the latter was successful and diol **10** could be converted into the corresponding dioxo derivative **11** (Figure 3). Starting from 0.50 g of diol **10**, according to the literature [32], a loading of 1 mol% for each catalyst was enough to afford complete conversion into the diketone in toluene solution at 100 °C in 5 h. After work-up (quenching with water and extraction), the crude residue was submitted directly to the following step of oxidative cleavage.

2.4. *Oxidative Cleavage of 9,10-Dioxostearic Acid* (**11**) *to Azelaic* (**2**) *and Pelargonic Acid* (**3**)

For the final step of the synthetic procedure, we investigated the Baeyer–Villiger (BV) oxidation of diketone **11**, to prepare the corresponding anhydride **12**, and hydrolyze it to azelaic (**2**) and pelargonic acid (**3**). A wide range of oxidants has been employed for the BV reaction, including mineral and organic peracids. Hydrogen peroxide can be used if suitably activated by a catalyst, or in the presence of a strong acid, or even in alkaline conditions [33,34]. α-Diketones react readily with BV reagents: in inert solvents anhydrides are formed, while in alkaline or acidic media simple carboxylic acids are generally produced in good yields [35]. In 1930 Böeseken et al. [36] prepared 9,10-diketostearic acid (**11**) by oxidation of 9-octadecynoic acid with 70% nitric acid at 10–25% yield and submitted it to the reaction with 15% excess peracetic acid in acetic acid for one day. They obtained the quantitative conversion of the diketone into a mixture of acids **2** and **3**.

Thus, we first considered the possibility to perform the BV oxidation of compound **11** with the corresponding peroxycarboxylic acid produced by lipase-mediated perhydrolysis in the presence of H_2O_2, using for preliminary experiments a sample of **11** isolated and purified by column chromatography. We treated dioxostearic acid **11** (50 mg) with 1.6 mol of H_2O_2 per mol of dioxostearic acid in the presence of 2.5 $mg{\cdot}mL^{-1}$ Novozyme 435 in toluene (2 mL), the same solvent employed for the preceding oxidation. The cleavage was complete after 3 h at 30 °C, to give a mixture of 96% pelargonic and azelaic acids, with tiny quantities of octanoic (0.3%), stearic (0.4%), and palmitic acid (2%), with other minor components (GC/MS). To our surprise, when the blank reaction was carried out in parallel in identical conditions without the presence of the enzyme, the same oxidative cleavage was observed, affording a mixture containing acids **2** and **3**, besides 51% of intermediate anhydride **12** (^1H NMR). The presence of intermediate **12** was highlighted by NMR spectroscopy. The ^{13}C NMR spectrum of the crude reaction mixture showed the presence of two singlets at 169.73 and 169.67 ppm for the carboxylic carbon atoms of the anhydride, next to those around 180 ppm which belong to acids **2** and **3** and to the COOH group of compound **12.** In the ^{1}H NMR spectrum, the triplet of the CH_2 groups linked to the CO-O-CO moiety is at 2.44 ppm, a little more deshielded than the triplet of the CH_2 beside COOH in compounds **2** and **3** and in the anhydride itself, occurring at 2.35 ppm. The ^{1}H and ^{13}C NMR spectra of anhydride **12** are not known in the literature, and the spectroscopic data of lauric anhydride reported in [37] were used as reference data.

The reaction was repeated in acetonitrile and only 11% of anhydride was found in the final mixture. When the oxidation was carried out in toluene and 2 M H_2SO_4 was added to the reaction mixture during the workup procedure, after having decomposed peroxy species with $NaHSO_3$ saturated solution, complete hydrolysis of intermediate **12** was obtained.

After investigation of every single step, the whole procedure was performed starting from 2 g of commercial oleic acid. Acidic hydrolysis was performed soon after epoxidation in a one-pot procedure, to afford diol derivative **10** as a pure compound at 70% isolated yield by filtration of the first crop of crystalline material and recovery of other product by further treatment of the mother liquors. The oxidation to dioxoderivative **11** gave a crude compound (75% purity by GC/MS) that was submitted directly to the last step of oxidative cleavage in toluene with only 35% H_2O_2, to provide a mixture of azelaic and pelargonic acids. Diacid **2** was recovered following a procedure, which had been already described in the literature [16], and based on the solubility of compound **2** in hot water. Repartition between ethyl acetate and hot water afforded an aqueous phase from which azelaic acid crystallised upon cooling. After three extraction cycles, diacid **2** could be recovered as a pure compound in 73% yield. Pelargonic acid **3** was isolated from the organic phase at 77% isolation yields, showing 91% chemical purity (GC/MS). The separation of diacid **2** from compound **3** was also investigated by using column chromatography, eluting with hexane–EtOAc mixtures with an increasing amount of the more polar solvent (see Supplementary Material), affording pure **2** and **3** in slightly higher isolation yields (81% and 84%, respectively).

3. Discussion

Ozonolysis of alkene bonds is a useful chemical transformation which is employed not only at the laboratory level but also at industrial scale for the rapid and effective oxidative cleavage of C=C double bonds [38]. The primary concern related to ozonolysis chemistry is represented by the serious safety issues connected with the reaction, and in particular with the explosive hazard due to the instability of intermediate ozonides. The present industrial production of azelaic acid is entirely based on ozonolysis of oleic acid, being the global azelaic acid market valued at 94 million USD in 2017 and expected to reach 140 million USD by 2025 [39]. The market is mainly driven by growing demand for plastics and lubricants, which hold above 70% of global azelaic acid consumption.

The growing attention towards the development of safer and more environmentally friendly production technologies has stimulated the investigation of alternative methods for the conversion of oleic acid into azelaic acid. Suitable references have been reported in the Introduction. We gave a contribution to this search by investigating a chemo-enzymatic approach to achieve the target oxidative scission.

We decided to use the in-situ peroxidation of oleic acid **4** by lipase-mediated perhydrolysis in the presence of hydrogen peroxide 35% as a safe procedure to afford the peroxycarboxylic acid needed to promote the epoxidation step at the beginning of the synthetic sequence. Oleic acid itself undergoes the conversion into the reactive peroxy species, so it is possible to avoid the use of an additional carboxylic acid that would remain in the reaction mixture as a by-product to be removed from the desired final compound. The advantage of the proposed procedure is also that storage and manipulation of peracid are avoided: it is generated in the reaction medium, and the excess is destroyed at the end of its use. The best biocatalyst for this reaction is Novozyme 435 which has the advantage of being an immobilized form of *C. Antarctica* B that can be recovered and re-used. Preliminary experiments were performed starting from 1 g of oleic acid, and recovering the enzyme by filtration and washing with water and acetonitrile. The enzyme was kept at 4 °C for 18 h and re-used in a subsequent reaction. After four runs the conversion of oleic acid (GC/MS analysis) into the epoxide was 78%. The evaluation of the enzyme reusability in these experiments can only be used as a first orientation because it is influenced by the effects of manipulation and storage on the enzyme support, overlapping those due to hydrogen peroxide and peracid. Long-term performance of the enzyme should be best studied under continuous process conditions, as suggested by recent literature [40]. Very positive results on the stability of this lipase in this type of reaction have been obtained using both a packed-bed reactor [41] and a continuous stirred tank reactor [40]. This kind of investigation is now in progress in our research group.

We considered also the possibility to avoid the isolation and purification of some of the intermediates of our procedure to reduce quantities of waste, solvents and separation aids. We chose the solvent of the epoxidation reaction to telescope the first two steps of the procedure, and perform the acid-catalyzed opening of the oxirane ring without isolation of epoxide **9**. Diol derivative **10** could be obtained as a pure crystalline compound by crystallization from acetonitrile, after quenching the peroxy species with $NaHSO_3$ and promoting epoxide cleavage with catalytic H_2SO_4 aqueous solution. No column chromatography was needed, thus favoring the isolation yield of diol **10** and limiting further use of solvents. Even the purification of dioxostearic acid **11** could be avoided, and the raw material was submitted directly to the following step to afford azelaic and pelargonic acid. The use of toluene as a solvent for both the diol oxidation and the final oxidative cleavage reaction will be useful in the future for developing the procedure in continuous flow mode.

For the last step of the whole sequence, we discovered the unexpected capability of H_2O_2 35% w/w to promote the oxidative cleavage of diketone **11** in organic solvents, either toluene or acetonitrile, at 30 °C, without the addition of any catalyst. H_2O_2 is considered as a green oxidant, generating water as a by-product. It is safely stored and transported, and easily available on the market at a cheap price.

4. Materials and Methods

4.1. General Methods

Chemicals and solvents were purchased from Merck (Merck Life Science S.r.l., Milan, Italy) and used without further purification. Trimethylsilyldiazomethane 10% solution in hexane (TCI Europe N.V.) was purchased from Zentek Srl (Milan, Italy). Novozyme 435 (Novozymes) was purchased from Strem Chemicals Inc. (Bischheim, France). TLC analyses were performed on Macherey Nagel pre-coated TLS sheets Polygram® SIL G/UV$_{254}$ purchased from Chimikart s.r.l. (Naples, Italy). ^{1}H and ^{13}C NMR spectra were recorded on a 400 or 500 MHz spectrometer in $CDCl_3$ solution at r.t. The chemical shift scale was based on internal tetramethylsilane. GC/MS analyses were performed using an HP-5MS column (30 m × 0.25 mm × 0.25 μm, Agilent Technologies Italia Spa, Cernusco sul Naviglio, Italy). The following temperature program was employed: 50 °C/10 °C min^{-1}/250 °C (5 min)/50 °C min^{-1}/300 °C (10 min). The samples for GC/MS were treated with MeOH and trimethylsilyldiazomethane 10% in hexane, to derivatize carboxylic acids by transformation into the respective methyl esters.

4.2. *One-Pot Two-Step Synthesis of Threo-9,10-Didhydroxystearic Acid* (**10**)

A suspension of Novozyme 435 (240 mg) in acetonitrile (48 mL), containing oleic acid (2.2 g, 91% purity, 7.1 mmol) and 35% H_2O_2 w/w (1.1 mL, 12.8 mmol) was shaken in an orbital shaker (160 rpm, 50 °C) for 5 h. The enzyme was removed by filtration, washed with acetonitrile, and stored at 4 °C to be re-used. A saturated solution of $NaHSO_3$ (2 mL) was added to the filtrate, followed by the addition of 2 M H_2SO_4 (963 μL). After 15 h at room temperature, diol **10** was recovered by filtration (1.62 g, 72%, sum of two crystallization crops). ^{1}H NMR (CD_3OD, 400 MHz) [42]: δ = 3.45–3.35 (2H, m, 2CHOH), 2.29 (2H, t with J = 7.4 Hz, CH_2COOH), 1.70–1.15 (26H, m, 13 CH_2), 0.97–0.82 (3H, m, CH_3). ^{13}C NMR (CD_3OD, 100.6 MHz) [42]: δ = 177.6, 75.29, 75.26, 35.0, 34.0, 33.9, 33.0, 30.8, 30.7, 30.6, 30.42, 30.37, 30.2, 27.04, 26.96, 26.1, 23.7, 14.4. GC/MS (EI) as a methyl ester, obtained by treatment with MeOH and trimethylsilyldiazomethane 10% in hexane, t_r = 23.85 min: m/z (%) = 294 (M^+ – 36, 1), 187 (48), 155 (100), 138 (30).

4.3. Oxidation of 9,10-Dihydroxystearic Acid (9) to 9,10-Dioxostearic Acid (10)

A mixture of diol **10** (1.55 g, 4.9 mmol), $Fe(NO_3)_3 \cdot 9\ H_2O$ (20 mg, 0.049 mmol), TEMPO (8.0 mg, 0.049 mmol), and NaCl (3 mg, 0.049 mmol) in toluene (45 mL) was stirred at 100 °C for 4–5 h. The reaction mixture was poured into water and extracted with ethyl acetate. The organic phase was dried and concentrated under reduced pressure to give the crude dioxo derivative **11** (1.83 g, 75% purity by GC/MS analysis, estimated content of compound **11** 1.37 g) which was employed in the successive step without further purification. ^1H NMR ($CDCl_3$, 400 MHz) [43]: δ = 2.72 (4H, t with *J* = 7.3 Hz, $2CH_2CO$), 2.35 (2H, t with *J* = 7.4 Hz, CH_2COOH), 1.69–1.49 (6H, m, 3 CH_2), 1.40–1.20 (16H, m, 8 CH_2), 0.92–0.84 (3H, m, CH_3). ^{13}C NMR ($CDCl_3$, 100.6 MHz) [43]: δ 200.3, 200.2, 180.0, 36.2, 36.1, 34.2, 31.9, 29.4, 29.3, 29.2, 29.1, 29.04, 28.96, 24.2, 23.2, 23.1, 22.8, 14.2. GC/MS (EI) as a methyl ester, obtained by treatment with MeOH and trimethylsilyldiazomethane 10% in hexane, t_r = 22.48 min: *m/z* (%) = 326 (M^+, 1), 295 (5), 185 (100), 141 (54).

4.4. Oxidative Cleavage of 9,10-Dioxostearic Acid (11) to Azelaic (2) and Pelargonic Acid (3)

A mixture of crude dioxo derivative **11** (1.75 g, 75% purity, estimated content of compound **11** 4.21 mmol) and 35% H_2O_2 w/w (579 µL, 6.73 mmol) in toluene (35 mL) was stirred at 30 °C for 3 h. A saturated solution of $NaHSO_3$ (750 µL) was added, followed by the addition of H_2SO_4 2 M till pH = 2. The reaction mixture was then extracted with ethyl acetate. The organic phase was dried and concentrated under reduced pressure to give a crude mixture containing 93% (GC/MS) of acids **2** and **3**, which was heated at 50 °C for 1 h in a 1:1 mixture of EtOAc and water. Water was separated, and diacid **2** crystallized upon cooling. Other two extractions of the organic phase with hot water allowed the isolation of diacid **2** as a pure compound. Pelargonic acid **3** was isolated from the organic phase showing 91% chemical purity (GC/MS).

Azelaic acid (**2**): 578 mg (73%, >99% chemical purity by GC/MS and NMR); ^1H NMR ($CDCl_3$, 400 MHz) [44]: δ = 2.35 (4H, t with *J* = 7.4 Hz, $2CH_2COOH$), 1.75–1.55 (4H, m, $2CH_2$), 1.4–1.2 (6H, m, $3CH_2$). ^{13}C NMR ($CDCl_3$, 100.6 MHz) [44]: δ = 180.2, 34.2, 29.0, 28.9, 24.3. GC/MS (EI) as a methyl ester, obtained by treatment with MeOH and trimethylsilyldiazomethane 10% in hexane, t_r = 13.9 min: *m/z* (%) = 185 (M^+ - 31, 55), 152 (100), 143 (47), 111 (63).

Pelargonic acid (**3**): 512 mg (77%, 91% chemical purity by GC/MS); ^1H NMR ($CDCl_3$, 400 MHz) [45]: δ = 2.35 (2H, t with *J* = 7.5Hz, CH_2COOH), 1.75–1.55 (2H, m, CH_2), 1.4–1.2 (10H, m, $5CH_2$), 0.80–0.95 (3H, m, CH_3). ^{13}C NMR ($CDCl_3$, 100.6 MHz) [45]: δ = 180.5, 34.2, 31.9, 29.3, 29.20, 29.18, 24.8, 22.7, 14.2. GC/MS (EI) as a methyl ester, obtained by treatment with MeOH and trimethylsilyldiazomethane 10% in hexane, t_r = 9.33 min: *m/z* (%) = 172 (M^+, 0.5), 141 (15), 129 (18), 87 (45), 74 (100).

5. Conclusions

The chemo-enzymatic conversion of oleic acid into azelaic and pelargonic acids herein described represents a sustainable alternative to ozonolysis, currently employed at the industrial scale. Azelaic acid can be produced in high chemical purity in 44% isolation yield after three steps, avoiding column chromatography purifications. Intermediate diol **10** and final azelaic acid **2** are purified by crystallization from acetonitrile and water, respectively. The procedure shows some valuable aspects, even if it is not a one-pot process, as those using H_2O_2 and tungsten derivatives already known in the literature [13–18].

The reagents of the three steps are: (i) H_2O_2 35% for the epoxidation of acid **4** and the oxidative cleavage of diketone **11**, and atmospheric oxygen for diol **10** oxidation, both producing H_2O as a by-product; (ii) H_2O for the acid-catalyzed hydrolysis of both epoxide **9** and anhydride **12**, generating no side-product, being fully incorporated in the reacting products. The organic solvents used during the reactions are limited to acetonitrile and toluene; water and ethyl acetate are employed for quenching and separation procedures. The final oxidative cleavage of dioxo derivative **11** occurs in mild conditions and generates a very tiny quantity of oxidized impurities, thus increasing the economic value of the process, and reducing the complexity and cost of final azelaic acid purification. Hydrogen peroxide is itself very effective in promoting the cleavage with no need for catalysts or harsh acidic or alkaline conditions and generating water as a side-product. The reaction medium can be either toluene or acetonitrile. The use of an enzymatic method to produce in situ H_2O_2 will be considered for further development of the process.

Studies are now in progress to apply this synthetic procedure to soapstock recovered from the neutralization step during vegetable seed oil refining at Oleificio Zucchi. A pre-treatment step has to be added where the lipase-mediated hydrolysis of the triglycerides, which are inevitably present in this by-product, is carried out.

A further development of the process will also be the optimization of the entire sequence in continuous flow mode, taking advantage of the fact the enzyme employed for the generation of the key peroxy species is already marketed in immobilized form. Besides an expected higher productivity value and an advantage for the process scalability study, the use of a continuous flow reactor will most likely increase the stability of lipase, and allow for a more reliable evaluation of lipase reusability.

Supplementary Materials: The following are available online, Table S1: Effect of H_2O_2/oleic acid molar ratio and temperature on the chemo-enzymatic epoxidation of oleic acid, Table S2: Effect of the amount of lipase and solvent on the chemo-enzymatic epoxidation of oleic acid, Table S3: Effect of the amount of aqueous 2M sulfuric acid on the ring-opening of epoxystearic acid **9**, Characterization of 8-((2*SR*,3*RS*)-3-octyloxiran-2-yl)octanoic acid (**9**), Recovery and re-use of Novozyme 435, Procedure for the oxidative cleavage of 9,10-dioxostearic acid (**11**) to give a mixture of 9-(nonanoyloxy)-9-oxononanoic acid (**12**), azelaic (**2**), and pelargonic acid (**3**), Column chromatography separation of azelaic (**2**) and pelargonic acid (**3**).

Author Contributions: Conceptualization, E.B. and D.T.; methodology, E.B., D.T., and G.D.L.; experimental procedure, M.V., M.C.G., F.T., and D.C.; characterization, F.G.G., M.V., M.C.G., and G.D.L.; data analysis, F.T., D.C., E.B.; writing—original draft preparation, E.B., F.T., D.C.; writing—review and editing, E.B., D.T. All authors have read and agreed to the published version of the manuscript.

Funding: This research was funded by Fondazione Cariplo—INNOVHUB, project SOAVE (Seed and vegetable Oils Active Valorization through Enzymes), grant number 2017-1015.

Conflicts of Interest: The authors declare no conflict of interest.

References

1. Available online: https://sustainabledevelopment.un.org/post2015/transformingourworld (accessed on 27 March 2020).
2. Soutelo-Maria, A.; Dubois, J.-L.; Couturier, J.-L.; Giancarlo Cravotto, G. Oxidative Cleavage of Fatty Acid Derivatives for Monomer Synthesis. *Catalysts* **2018**, *8*, 464. [CrossRef]
3. Fraile, J.M.; García, J.I.; Herrerías, C.I.; Pires, E. Transformations for the Valorization of Fatty Acid Derivatives. *Synthesis* **2017**, *49*, 1444–1460. [CrossRef]
4. Song, J.-W.; Seo, J.-H.; Oh, D.-K.; Bornscheuer, U.T.; Park, J.-B. Design and engineering of whole-cell biocatalytic cascades for the valorization of fatty acids. *Catal. Sci. Technol.* **2020**, *10*, 46–64. [CrossRef]
5. Dubois, J.-L. Arkema's Integrated Plant-Based Factories. In *Chemicals and Fuels from Bio-Based Building Block*; Cavani, F., Albonetti, S., Basile, F., Gandini, A., Eds.; Wiley-VCH Verlag GmbH & Co. KGaA: Weinheim, Germany, 2016; Chapter 20; pp. 545–548.
6. Ackman, R.G.; Retson, M.E.; Gallay, L.R.; Vandenheuvel, F.A. Ozonolysis of unsaturated fatty acids. i. Ozonolysis of oleic acid. *Can. J. Chem.* **1961**, *39*, 1956–1963. [CrossRef]
7. Brydson, J.A. Polyamides and Polyimides. In *Plastics Materials*, 7th ed.; Butterworth Heinemann: Oxford, UK, 1999; Chapter 18; pp. 478–530.
8. Available online: https://www.drugs.com/mtm/azelaic-acid-topical.html (accessed on 31 December 2019).
9. Sasmaz, S.; Arican, O. Comparison of azelaic acid and anthralin for the therapy of patchy alopecia areata: A pilot study. *Am. J. Clin. Dermatol.* **2005**, *6*, 403–406. [CrossRef]
10. Pohanish, R.P. *Sittig's Handbook of Pesticides and Agricultural Chemicals*, 2nd ed.; William Andrew: Norwich, NY, USA, 2015.

11. Anneken, D.J.; Both, S.; Christoph, R.; Fieg, G.; Steinberner, U.; Westfechtel, A. Fatty acids. In *Ullmann's Encyclopedia of Industrial Chemistry*; Wiley-VCH Verlag GmbH & Co. KGaA: Weinheim, Germany, 2000; pp. 73–116.
12. Kerenkan, A.E.; Béland, F.; Do, T.-O. Chemically catalyzed oxidative cleavage of unsaturated fatty acids and their derivatives into valuable products for industrial applications: A review and perspective. *Catal. Sci. Technol.* **2016**, *6*, 971–987. [CrossRef]
13. Antonelli, E.; D'Aloisio, R.; Gambaro, M.; Fiorani, T.; Venturello, C. Efficient Oxidative Cleavage of Olefins to Carboxylic Acids with Hydrogen Peroxide Catalyzed by MethyltrioctylammoniumTetrakis (oxodiperoxotungsto) phosphate (3-) under Two-Phase Conditions. Synthetic Aspects and Investigation of the Reaction Course. *J. Org. Chem.* **1998**, *63*, 7190–7206. [CrossRef]
14. Li, X.; Syong, J.C.P.; Zhang, Y. Sodium stannate promoted double bond cleavage of oleic acid by hydrogen peroxide over a heterogeneous WO_3 catalyst. *Green Chem.* **2018**, *20*, 3619–3624. [CrossRef]
15. Enferadi-Kerenkan, A.; Gandon, A.; Do, T.O. Novel tetra-propyl/butylammonium encapsulated Keggin-type polyoxotungstates: Synthesis, structural characterization, and catalytic capability in oxidative cleavage of unsaturated fatty acids. *Dalton Trans.* **2018**, *47*, 1214–1222. [CrossRef]
16. Benessere, V.; Cucciolito, M.E.; De Santis, A.; Di Serio, M.; Esposito, R.; Ruffo, F.; Turco, R. Sustainable Process for Production of Azelaic Acid Through Oxidative Cleavage of Oleic Acid. *J. Am. Oil. Chem. Soc.* **2015**, *92*, 1701–1707. [CrossRef]
17. Kadyrov, R.; Hackenberger, D. Oxidative Cleavage of Long Chain Olefins to Carboxylic Acids with Hydrogen Peroxide. *Top. Catal.* **2014**, *57*, 1366–1371. [CrossRef]
18. Godard, A.; De Caro, P.; Thiebaud-Roux, S.; Vedrenne, E.; Mouloungui, Z. New Environmentally Friendly Oxidative Scission of Oleic Acid into Azelaic Acid and Pelargonic Acid. *J. Am. Oil Chem. Soc.* **2013**, *90*, 133–140. [CrossRef]
19. Song, J.-W.; Jeon, E.-Y.; Song, D.-H.; Jang, H.-Y.; Bornscheuer, U.T.; Oh, D.-K.; Park, J.-B. Multistep Enzymatic Synthesis of Long-Chain α,ω-Dicarboxylic and ω-Hydroxycarboxylic Acids from Renewable Fatty Acids and Plant Oils. *Angew. Chem. Int. Ed.* **2013**, *52*, 2534–2537. [CrossRef] [PubMed]
20. Song, J.-W.; Lee, J.-H.; Bornscheuer, U.T.; Park, J.-B. Microbial Synthesis of Medium-Chain α,ω Dicarboxylic Acids and ω-Aminocarboxylic Acids from Renewable Long-Chain Fatty Acids. *Adv. Synth. Catal.* **2014**, *356*, 1782–1788. [CrossRef]
21. Cha, H.J.; Seo, E.-J.; Song, J.-W.; Jo, H.-J.; Kumar, A.R.; Park, J.-B. Simultaneous Enzyme/Whole-Cell Biotransformation of C18 Ricinoleic Acid into (R)-3-Hydroxynonanoic Acid, 9-Hydroxynonanoic Acid, and 1,9-Nonanedioic Acid. *Adv. Synth. Catal.* **2018**, *360*, 696–703. [CrossRef]
22. Otte, K.B.; Kittelberger, J.; Kirtz, M.; Nestl, B.M.; Hauer, B. Whole-Cell One-Pot Biosynthesis of Azelaic Acid. *ChemCatChem* **2014**, *6*, 1003–1009. [CrossRef]
23. Otte, K.B.; Kirtz, M.; Nestl, B.M.; Hauer, B. Synthesis of 9-Oxononanoic Acid, a Precursor for Biopolymers. *ChemSusChem* **2013**, *6*, 2149–2156. [CrossRef]
24. Kim, J.-Y.; Jun, M.-W.; Seong, Y.-J.; Park, H.; Ahn, J.; Park, Y.-C. Direct Biotransformation of Nonanoic Acid and Its Esters to Azelaic Acid by Whole Cell Biocatalyst of *Candida tropicalis*. *ACS Sustain. Chem. Eng.* **2019**, *7*, 17958–17966. [CrossRef]
25. Oleificio Zucchi S.p.A. *(Cremona), Internal Analysis: 55–60% Oleic Acid, 30–35% Linoleic Acid*; Oleificio Zucchi S.p.A.: Cremona, Italy.
26. Koppireddi, S.; Seo, J.-H.; Jeon, E.-Y.; Chowdhury, P.S.; Jang, H.-Y.; Park, J.-B.; Kwona, Y.-U. Combined Biocatalytic and Chemical Transformations of Oleic Acid to ω-Hydroxynonanoic Acid and α,ω-Nonanedioic Acid. *Adv. Synth. Catal.* **2016**, *358*, 3084–3092. [CrossRef]
27. Guicheret, B.; Bertholo, Y.; Blach, P.; Raoul, Y.; Métay, E.; Lemaire, M. A Two-Step Oxidative Cleavage of 1,2-Diol Fatty Esters into Acids or Nitriles by a Dehydrogenation–Oxidative Cleavage Sequence. *ChemSusChem* **2018**, *11*, 3431–3437. [CrossRef]
28. Activated Bleaching Composition. The Clorox Company (California, US). U.S. Patent EP253487, 4 June 1987.
29. Bjorkling, F.; Godtfredsen, S.-E.; Kirk, O. Lipase-mediated Formation of Peroxycarboxylic Acids used in Catalytic Epoxidation of Alkenes. *J. Chem. Soc. Chem. Commun.* **1990**, 1301–1303. [CrossRef]
30. Warwel, S.; Klaas, M.R.G. Chemo-enzymatic epoxidation of unsaturated carboxylic acids. *J. Mol. Catal. B Enzym.* **1995**, *1*, 29–35. [CrossRef]

31. Milchert, E.; Malarczyk, K.; Kłos, M. Technological Aspects of Chemoenzymatic Epoxidation of Fatty Acids, Fatty Acid Esters and Vegetable Oils: A Review. *Molecules* **2015**, *20*, 21481–21493. [CrossRef] [PubMed]
32. Ma, S.; Liu, J.; Li, S.; Chen, B.; Cheng, J.; Kuang, J.; Liu, Y.; Wan, B.; Wang, Y.; Ye, J.; et al. Development of a General and Practical Iron Nitrate/TEMPO Catalyzed Aerobic Oxidation of Alcohols to Aldehydes/Ketones: Catalysis with Table Salt. *Adv. Synth. Catal.* **2011**, *353*, 1005–1017. [CrossRef]
33. ten Brink, G.-J.; Arends, I.W.C.E.; Sheldon, R.A. The Baeyer-Villiger Reaction: New Developments toward Greener Procedures. *Chem. Rev.* **2004**, *104*, 4105–4123. [CrossRef]
34. Krow, G.R. The Baeyer–VilligerOxidation of Ketones and Aldehydes. *Org. React.* **1993**, *43*, 251–809. [CrossRef]
35. Hassall, C.H. The Baeyer-Villiger Oxidation of Aldehydes and Ketones. *Org. React.* **1957**, *9*, 73–94. [CrossRef]
36. Böeseken, J.; Slooff, G. L'action de l'acide peracétique sur l'ac. dicéto-9.10-stéarique, le benzile, la β-naphtaquinoneet l'ortho-quinone simple. *Rec. Trav. Chim.* **1930**, *49*, 91–94. [CrossRef]
37. Baldwin, L.C.; Davis, M.C.; Hughes, A.M.; Lupton, D.V. Potential Vegetable-Based Diesel Fuels from Perkin Condensation of Furfuraldehyde and Fatty Acid Anhydrides. *J. Am. Oil. Chem. Soc.* **2019**, *96*, 571–583. [CrossRef]
38. Van Ornum, S.G.; Champeau, R.M.; Pariza, R. Ozonolysis Applications in Drug Synthesis. *Chem. Rev.* **2006**, *106*, 2990–3001. [CrossRef]
39. Available online: http://www.digitaljournal.com/pr/3836508 (accessed on 29 December 2019).
40. Meyer, J.; Holtmann, D.; Ansorge-Schumacherc, M.B.; Kraumed, M.; Drews, A. Development of a continuous process for the lipase-mediated synthesis of peracids. *Biochem. Eng. J.* **2017**, *118*, 34–40. [CrossRef]
41. Wiles, C.; Hammond, M.J.; Watts, P. The development and evaluation of a continuous flow process for the lipase-mediated oxidation of alkenes. *Beilstein J. Org. Chem.* **2009**, *5*. [CrossRef] [PubMed]
42. Awang, R.; Ahmad, S.; Kang, Y.B.; Ismail, R. Characterization of Dihydroxystearic Acid from Palm Oleic Acid. *J. Am. Oil Chem. Soc.* **2001**, *78*, 1249–1252. [CrossRef]
43. Kulik, A.; Martin, A.; Pohl, M.-M.; Fischer, C.; Köckritz, A. Insights into gold-catalyzed synthesis of azelaic acid. *Green Chem.* **2014**, *16*, 1799–1806. [CrossRef]
44. Schievano, E.; Morelato, E.; Facchin, C.; Mammi, S. Characterization of Markers of Botanical Origin and Other Compounds Extracted from Unifloral Honeys. *J. Agric. Food Chem.* **2013**, *61*, 1747–1755. [CrossRef]
45. Liu, Y.; Cornella, J.; Martin, R. Ni-Catalyzed Carboxylation of Unactivated Primary Alkyl Bromides and Sulfonates with CO_2. *J. Am. Chem. Soc.* **2014**, *136*, 11212–11215. [CrossRef]

Sample Availability: No samples are available from the authors.

Article

Mutagenesis and Adaptation of the Psychrotrophic Fungus *Chrysosporium pannorum* A-1 as a Method for Improving β-pinene Bioconversion

Mateusz Kutyła, Jan Fiedurek, Anna Gromada, Krzysztof Jędrzejewski and Mariusz Trytek *

Department of Industrial and Environmental Microbiology, Faculty of Biology and Biotechnology, Maria Curie-Skłodowska University, Akademicka 19, 20-033 Lublin, Poland; mateusz.kutyla@umcs.pl (M.K.); janek@umcs.pl (J.F.); anna.gromada@umcs.pl (A.G.); krzysztof.jedrzejewski@umcs.pl (K.J.)

* Correspondence: mariusz.trytek@umcs.pl or mtrytek1@o2.pl; Tel.: +48-81-537-5958

Academic Editor: Josefina Aleu
Received: 12 May 2020; Accepted: 1 June 2020; Published: 2 June 2020

Abstract: Mutagenesis and adaptation of the psychrotrophic fungus *Chrysosporium pannorum* A-1 to the toxic substrate β-pinene were used to obtain a biocatalyst with increased resistance to this terpene and improved bioconversion properties. Mutants of the parental strain were induced with UV light and *N*-methyl-*N′*-nitro-*N*-nitrosoguanidine. Mutants resistant to β-pinene were isolated using agar plates with a linear gradient of substrate concentrations. Active mutants were selected based on their general metabolic activity (GMA) expressed as oxygen consumption rate. Compared to the parental strain, the most active mutant showed an enhanced biotransformation ability to convert β-pinene to *trans*-pinocarveol (315 mg per g of dry mycelium), a 4.3-fold greater biocatalytic activity, and a higher resistance to H_2O_2-induced oxidative stress. Biotransformation using adapted mutants yielded twice as much *trans*-pinocarveol as the reaction catalyzed by non-adapted mutants. The results indicate that mutagenesis and adaptation of *C. pannorum* A-1 is an effective method of enhancing β-bioconversion of terpenes.

Keywords: adaptation; biotransformation; UV/NTG mutagenesis; psychrotrophs; terpenes

---

1. Introduction

Terpenes are valuable natural substrates commonly used in the production of fine chemicals. Turpentine, obtained from biomass and also as a side product of the softwood industry, is rich in monoterpenes, such as α-pinene and β-pinene, which are widely used as raw materials in the synthesis of flavors, fragrances, and pharmaceutical compounds [1,2]. Oxidative transformation of abundant and low-priced monoterpenes holds considerable potential for the production of a wide variety of different terpenoid derivatives which are difficult and often costly to obtain directly from plant species. Oxygenated monoterpenes are the main flavor and fragrance impact molecules of essential oils. Many of them have beneficial effects on health [3–5], including anti-inflammatory activity [6]. However, their high volatility, poor solubility in aqueous media, and natural toxicity to microorganisms still limit the use of these compounds in applied biocatalysis. These limitations can be overcome by slow/gradual addition of terpene into the reaction medium, use of biphasic systems containing biocompatible organic solvents or a co-solvent, and in situ product removal techniques [7–9].

β-Pinene is commercially available at US$ 36 per kg, whereas the price of its oxygenated derivative, *trans*-pinocarveol, can be as high as US$ 134,000 per kg [10,11]. There are data regarding the antimicrobial properties of β-pinene, used either as a purified chemical or as a constituent of essential oils [12,13]. Some studies have also demonstrated that β-pinene and *trans*-pinocarveol, contained in the oils, show various biological effects [14–18].

Some microbial isolates have been found capable of transforming β-pinene into α-terpineol [19]. However, the rate and direction of the conversion are often inadequate for the process to be economically viable [20].

Strain developers have searched for improved strains among random survivors of mutagenesis. Even though the outcome of classical mutagenesis is difficult to predict, and the selection of mutants is always phenotypic, it still remains a fact that many producer strains with enhanced productivity currently used in industrial processes have been generated by random mutagenesis [21–23]. In contrast to genetic engineering methods, classical mutagenesis facilitates obtaining highly efficient mutants in an easy and rapid way without specialized knowledge about microbe genomes. To obtain the required mutants, experimenters have enriched cell populations by culturing them in special environmental conditions, toxic to most cell types but less toxic or non-toxic to a desired minority of cells [24–28]. Induced mutation has been successfully used in a number of microbial processes (other than biotransformation) employed in the production of useful end products [27–29].

Organic-solvent-tolerant mutants of bacteria can be selected using mutagen treatments or genetically engineered from solvent-sensitive parental strains [30–32]. Microbial tolerance to organic solvents can be improved by transforming cloned genes which encode various proteins involved in this tolerance, located in the cytoplasm or the inner or outer membrane [33,34]. Overexpression of genes which confer tolerance to specific organic solvents results in enhanced tolerances, which can be put to practical use [35].

Mutants resistant to a particular type of abiotic stress were better adapted to other stress conditions, as expressed by their enzymatic activities. The ability of one stress condition to provide protection against other stresses is referred to as cross-protection [36]. Adaptation processes are used in many branches of industrial biotechnology, including biotransformation of chemical compounds [37–39], biofuel production [40,41], and polymer synthesis [42].

In a previous paper, we demonstrated that the psychrotrophic fungus *Chrysosporium pannorum* A-1 showed promise for the biotransformation of (1S)-(−)-α-pinene because it could be used at 20°C. This is an advantage in bioprocess involving volatile terpenes. Thus far, it has also given the best yields of verbenol and verbenone (722 and 176 mg/L, respectively) among the microorganisms studied [43].

In the present experiments, mutagenesis (UV irradiation and *N*-methyl-*N*′-nitro-*N*-nitrosoguanidine (NTG)) and adaptation to a toxic substrate (β-pinene) were used to select mutants of the psychrotrophic fungus *C. pannorum* A-1 characterized by an improved efficiency of biotransformation of β-pinene to the main product *trans*-pinocarveol.

2. Results and Discussion

Biotransformation of hydrophobic terpenes is limited by their toxicity to microbial biocatalysts. The metabolic activity of microorganisms, which is strongly dependent on environmental parameters, may also be affected by stressful conditions. It has been shown that the effectiveness of biotransformation largely relies on the interactions between the biocatalyst and environmental stressors, which may increase the yields of biotransformation products [44]. The influence of preincubation of the fungus *C. pannorum* A-1 under different stress conditions (organic solvents, medium pH, and temperatures) on its activity in oxidative bioconversion of α-pinene to verbenone and verbenol was examined in our previous study [44]. Since many genes are responsible for resistance to abiotic stresses, classical mutagenesis methods combined with adaptation can provide an alternative for fast generation of efficient mutants. Mutants characterized by resistance to different stress factors can be of potential use for biotransformation of toxic organic compounds. Classical mutagenesis is especially important when the metabolic pathways and the genome in the host strain have not yet been determined, as is the case with *C. pannorum*, the species used in the present study.

The lethal effect of UV used along with 0.01% NTG was studied by exposing *C. pannorum* A-1 to these mutagens for different time periods. The results show that the survival rate dropped significantly at a UV exposure time longer than 10 min and an NTG exposure time longer than 5 min. More precisely, it was found that 10 min of exposure to UV and 5 min of exposure to 0.01% NTG resulted in an approximately 9.6% survival rate, while 10 min of UV irradiation combined with 10–15 min of exposure to 0.01% NTG gave a survival rate of 1.6%. When exposure times were prolonged to 20 min for 0.01% NTG and 15 min for UV, the survival rate decreased dramatically to 0.37% (Table S1).

The gradient plate technique was found to be applicable to our mutant selection procedure, which uses β-pinene, a compound poorly soluble in water. We confirmed the linear concentration gradient of this substrate in agar plate regions (by Gas Chromatography (GC) analysis, data not shown). The gradient was consistent with the number of colonies growing in particular agar plate regions (Figure S1). A total of 137 mutants that grew well in the regions with the highest β-pinene concentration were isolated, and their general metabolic activity (GMA) was determined. Thirty mutants characterized by a higher GMA compared to non-treated strain were selected for further experiments. They were transferred to adaptive medium containing β-pinene and to non-adaptive malt agar medium (Figure S2).

Then, a second selection round was performed on the basis of the mutants' GMA. The results for the most active mutants are given in Figure 1 and oxygen uptake rates are shown in Table S2. In this selection round, twelve most active mutants were selected, which constituted 8.8% of all mutants. They were used for β-pinene biotransformation. Some of the mutants showed a substantially increased capacity to transform β-pinene to *trans*-pinocarveol (Figure 2). The most active non-adapted mutant (2–3) was characterized by a 2.4-fold increased biotransformation activity, whereas the average increase in activity among non-adapted mutants was 1.2-fold compared to the parental strain ($p < 0.05$) (ANOVA, Tukey's test). The efficiency of the biotransformation performed using mutants adapted to 1% β-pinene was on average 1.7-fold higher (per 1 g of dry weight of mycelium) than for non-adapted mutants. Large increases in biotransformation activity were observed for mutants: 2–6 (4.0-fold), 1–6 (2.4-fold), and 1–11 (1.8-fold). The most active mutant (1–15) was characterized by a 4.3-fold and 7-fold higher biotransformation yield in comparison to the adapted control and the non-adapted control, respectively ($p < 0.05$) (ANOVA, Tukey's test). The results are presented in Figure 2.

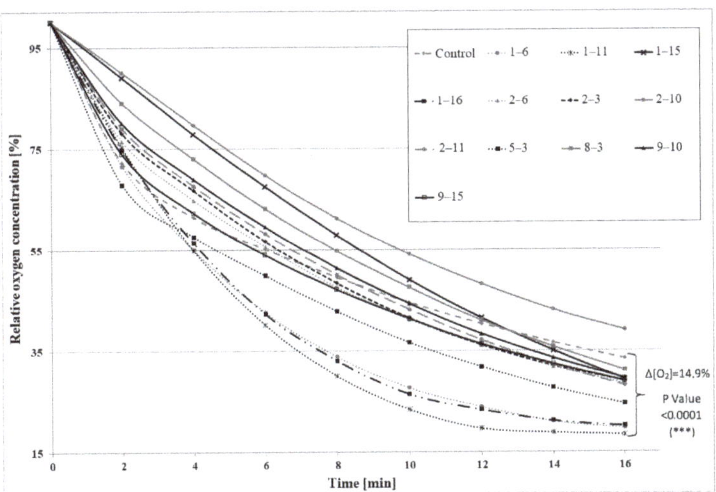

Figure 1. General metabolic activity (GMA) of 12 mutants of *C. pannorum* A-1 adapted to toxic β-pinene and selected for their high biotransformation ability. A non-treated wild strain was used as a control. Biomass concentration in each sample was 5 g dry mass per L.

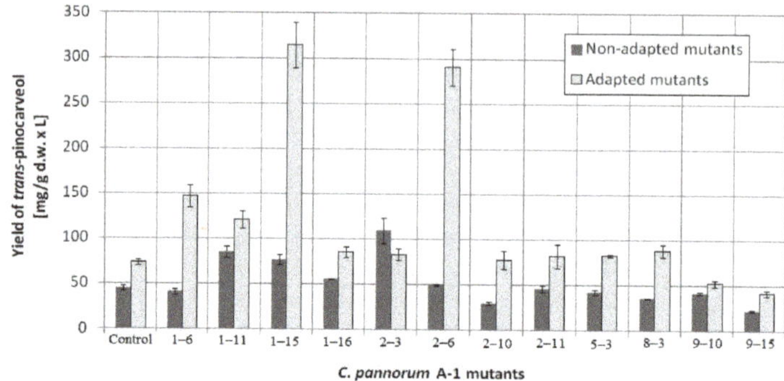

Figure 2. Effect of mutagenesis and adaptation to the substrate β-pinene on the biotransformation yield of 12 most active mutants of *C. pannorum* A-1. Non-treated parental strain and parental strain adapted to pinene were used as controls.

The main product of β-pinene biotransformation by the examined mutants was *trans*-pinocarveol, except for mutant 9–15, which accumulated substantial amounts of an unidentified product with a retention time of 20.1 min (the mass spectrum of this product is shown in Figure S3). This compound is probably produced by further metabolization of *trans*-pinocarveol, which might be catalyzed by a nonspecific enzyme induced during pinene biotransformation. This is more likely than not, as the fungal mutants neither grew on *trans*-pinocarveol as sole carbon source nor biotransformed it (data not shown). The fact that only one main product, with small total amounts (≤ 35%) of side-products (Figure 3), was produced is advantageous because usually biotransformation processes yield mixtures of different compounds which are then more difficult to separate.

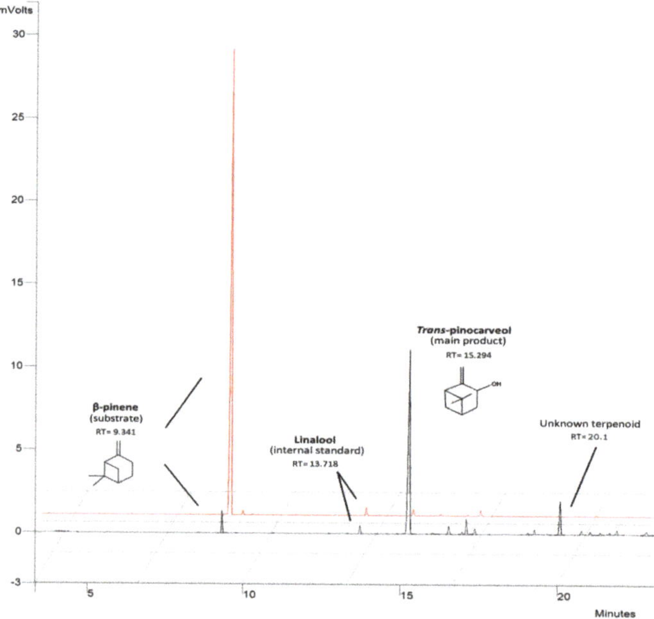

Figure 3. Gas Chromatography with Flame-Ionization Detection (GC-FID) chromatogram of main terpenoids obtained after 48 h of biotransformation of β-pinene by mutants of *C. pannorum* A-1. The initial substrate concentration was 1% (*v/v*). The red chromatogram refers to an abiotic control.

For the twelve most active mutants, β-pinene biotransformation efficiency was associated with the survivability of conidia (in the range from 1.6 to 9.6%) after mutagenesis (Table 1). The highest biotransformation efficiencies were obtained by the mutants which showed 9.6% conidia survivability. An increase in the biotransformation of α-pinene to verbenol using *Aspergillus niger* and *Penicillium* spp. after treatment with UV irradiation, colchicine, or ethyl methanesulfonate (EMS) was reported by Agrawal et al. [45]. The best results were obtained when UV irradiation was used: the biotransformation efficiency increased 15- and 8-fold, respectively, reaching 58.5 mg/g d.w. × L (mg of the product per gram of dry weight of the mycelium per liter aqueous phase) for *A. niger* and 19.5 mg/g d.w. × L for *Penicillium* spp. By contrast, both strains produced substantially lower quantities of verbenone for all the treatments used, relative to the wild-type strain. In addition, compared to our study, those authors obtained much lower (5.4–16.2-fold) product concentrations [45].

Table 1. Relationship between survivability of conidia obtained using mutagenesis and the efficiency of biotransformation of β-pinene to the main product *trans*-pinocarveol by mutants of *C. pannorum* A-1.

Variant of Mutagenesis	Survivability [%]	The Most Biotransformation-Active Mutants	Efficiency of Biotransformation [mg/g d.w. × L]
10 min UV + 5 min NTG	9.6	1–15 [1]	314.7 (±25.3)
		1–6 [1]	147.2 (±11.9)
		1–11 [1]	121.4 (±9.5)
		8–3 [1]	88.5 (±6.3)
		1–16 [1]	85.6 (±6.0)
10 min UV + 10 min NTG	1.6	2–6 [1]	290.8 (±20.4)
		2–3 [2]	109.4 (±14.2)
		2–11 [1]	82.2 (±13.1)
		2–10 [1]	77.8 (±10.1)
		9–10 [1]	52.6 (±3.7)

[1] adapted mutants; [2] non-adapted mutant. NTG – *N*-methyl-*N*′-nitro-*N*-nitrosoguanidine.

Phenomena associated with the adaptation of microbial cells to high concentrations of toxic substrates have been investigated in various studies [46–51]. For example, exposure of yeast cells to a stepwise increase in the level of ethanol stress has been shown to be an effective method of obtaining ethanol-tolerant yeast strains [46]. Compared with the parental strain, chemically mutagenized and spontaneous mutants exhibited increased acclimation and elevated growth rates when cultivated in sublethal ethanol concentrations, and they showed an increased survivability in lethal ethanol concentrations. Their higher tolerance to ethanol was due to the fact that they showed elevated glycerol production rates, which, in turn, was associated with an increase in the ratio of oxidized and reduced forms of nicotinamide adenine dinucleotide (NAD^+/NADH) in an ethanol-compromised cell, which stimulated the yeast cells' glycolytic activity [52].

Until now, no data have been published on the induction of psychrotrophic mutants using mutagenesis and adaptation to toxic substrates. There are also no reports on the ability of psychrotrophic fungi to biocatalyze the oxidation of β-pinene at temperatures below 25 °C. In this study, we attempted to fill the gap in previous research by establishing whether the mutants of the fungus *C. pannorum* A-1 had the biotechnological capability to catalyze an oxidation reaction at a low temperature and we also wanted to select mutants which produced high yields of high-value terpenoids from β-pinene. As far as we know, there are also no comprehensive reports regarding the effect of cell oxygen uptake on the biocatalytic activity of fungi. In our first experiments (Figures 1 and 2), no correlation was found between oxygen consumption by the fungal strains and the efficiency of the β-pinene biotransformation reaction estimated after 48 h. Therefore, the kinetic experiment was performed to analyze differences in the biocatalytic activities between the parental strain and mutant 1–6, which differed in oxygen uptake rate (k = 0.0271 and k = 0.0445, respectively) (Table S2). Attention was paid to some of the most important biocatalyst characteristics, such as resistance to process conditions and a short biotransformation time. An increase in biomass growth was observed during biotransformation on basal medium (BM) (Figure 4A). A higher biomass yield was produced by the parental strain ($p < 0.001$) (ANOVA, Tukey's test). On the rich BM medium, the metabolism of the parental strain was geared toward utilization of carbon, phosphorous and nitrogen sources (i.e., primary metabolism). In consequence, the higher biomass yield was accompanied by a higher *trans*-pinocarveol yield, compared to the mutant (Figure 5). However, in these conditions, the mutant adapted to β-pinene showed a shorter biotransformation time (36 h) in comparison to the parental strain (48–60 h) ($p < 0.05$) (ANOVA, Tukey's test). In the case of the mutant, a lower yield of *trans*-pinocarveol was obtained along with a higher substrate depletion (Figure 5B). This indicates that further metabolization of the product must have occurred.

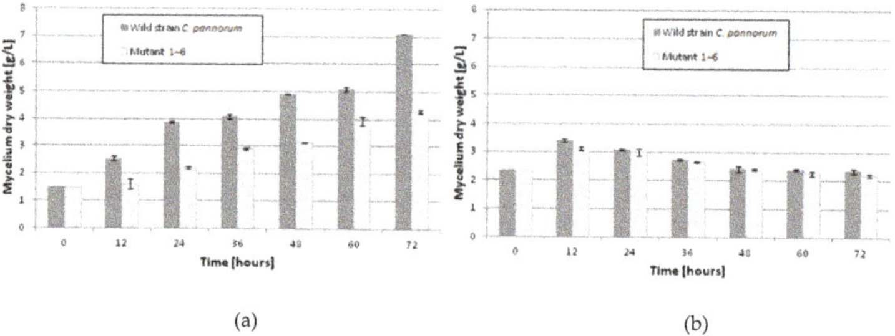

Figure 4. Cell growth rate of wild-type strain and mutant 1–6 of the fungus *C. pannorum* A-1 during biotransformation on rich basal medium (BM) (**a**) and 1% glucose medium containing phosphate buffer (**b**). Biotransformation conditions: temperature, 20 °C; initial β-pinene concentration, 1% (*v*/*v*). Bars represent the standard deviation of two independent samples.

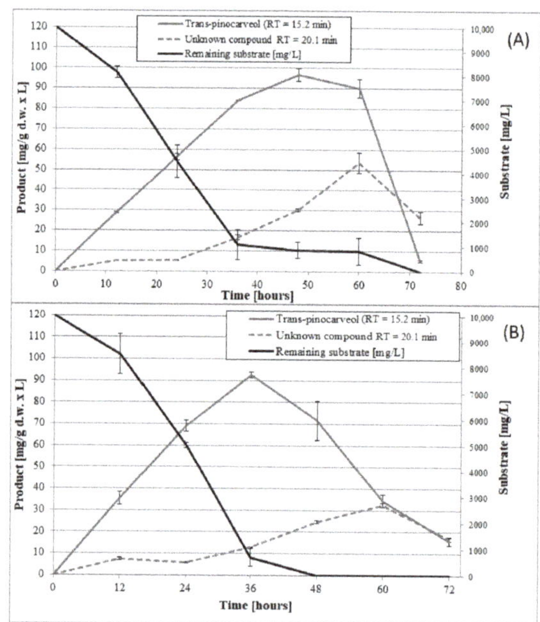

Figure 5. Time course of β-pinene biotransformation on basal medium (BM) by the wild-type strain (**A**) and mutant 1–6 (**B**) of the fungus *C. pannorum* A-1. An identical biomass concentration (1.5 g dry mass per L) was used in each sample. Bars represent the standard deviation of two independent samples.

When incubated in phosphate buffer, mutant 1–6 produced an about 1.4-fold higher yield of *trans*-pinocarveol from β-pinene compared to the parental strain (Figure 6) ($p < 0.05$) (ANOVA, Tukey's test). These two strains also differed in the optimum biotransformation time (depending on the main product), which varied from 48 to 72 h for the parental strain, and was 48 h for the mutant. The amount of dry mycelium of both strains decreased gradually over a period of 60 h after biotransformation on the buffer medium (Figure 4B), due to autolysis induced by nutrient deprivation.

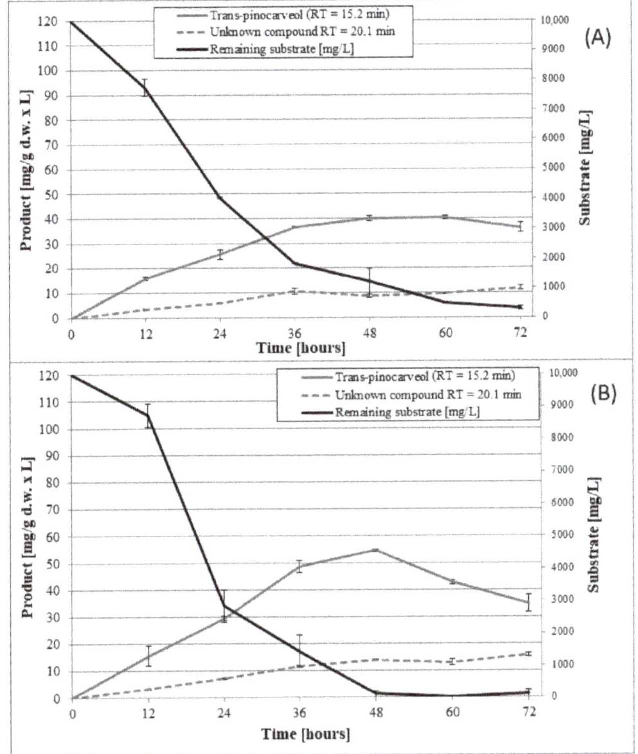

Figure 6. Time course of β-pinene biotransformation by the wild-type strain (**A**) and mutant 1–6 (**B**) of the fungus *C. pannorum* A-1 in phosphate buffer containing 1% glucose. An identical biomass concentration (2.35 g dry mass per L) was used in each sample. Bars represent the standard deviation of two independent tests.

The fact that the mutant produced higher yields of the unknown product (RT = 20.1 min) and caused higher substrate depletion (48 h) compared to the parental strain (72 h), ($p < 0.05$) (ANOVA, Tukey's test) confirmed the higher biocatalytic activity of the former. The results showed that biotransformation efficiency could be improved by using psychrotropic mutants characterized by higher oxygen consumption rates relative to the wild-type strain. Our results are consistent with the findings of Weber et al., who observed an appreciable increase in the rate of oxygen consumption by the mesophilic fungus *Cladosporium sphaerospermum* after addition of biodegradable substrates (i.e., benzyl alcohol, benzaldehyde, and catechol) [53].

A comparison of the biocatalytic efficiencies of the mutants adapted versus non-adapted to the substrate showed that the former produced higher yields of the biotransformation product (Figure 3). Cell adaptation to a solvent, substrate, and/or product has been found to be a successful strategy for reducing reagent inhibition in bioconversion systems [54,55]. Pre-delivery of low amounts of such compounds may result in increased synthesis of enzymes responsible for the transformation of xenobiotics [56]. Psychrotrophic fungi are adapted to survive in cold and very nutrient-poor environments, mainly due to the composition of their cell membranes and proteins. In addition, their enzymes are able to transform many xenobiotics into compounds that can be used as energy sources or ones that are less toxic to the cell [57–59]. The appearance of xenobiotics in the environment may cause stress to microorganisms, which, in turn, may induce adaptation mechanisms, such as changes in membrane composition or activation of genes responsible for the synthesis of chaperone proteins and catabolic enzymes, mainly cytochromes P450 [60,61]. Adaptation of microorganisms to a toxic substrate or induction with sublethal concentrations thereof often gives them an appropriate phospholipid profile [62,63]. In addition, bacterial cells (e.g., *Pseudomonas rhodesiae*) exposed to such harmful solvents as toluene and chloroform often undergo permeabilization, which increases biotransformation rates by improving the diffusion of products and substrates through the cell membrane [64]. Moreover, adaptation responses to one sub-lethal stress can lead to reduced susceptibility to a different stress (cross-protection or cross-adaptation) [65].

Since sublethal stresses in the β-pinene environment might induce stress-adaptive responses that could possibly make fungal cells resistant to other lethal stress factors [66], in the next experiment, we examined the resistance of *C. pannorum* A-1 strains to hydrogen peroxide. A remarkable increase in the decomposition of H_2O_2 was observed for mutant 1–6 relative to the wild strain. A 3-fold higher concentration of H_2O_2 (1.5% v/v) was required to inhibit the metabolic activity of the mutant compared to the wild strain (0.5% *v/v*), as shown by detection of H_2O_2 (undecomposed by catalase) in the medium after 1 h incubation (Table 2). This result was confirmed by measuring the GMA of the H_2O_2-stressed mycelia (pre-incubated with H_2O_2 at concentrations of 0.1–5% *v/v*). The examined concentrations of H_2O_2 above 0.1% caused a gradual fall in the rate of oxygen uptake by parental *C. pannorum* A-1 (Figure 7A). Incubation in the presence of 1.0% of H_2O_2 led to an almost total inhibition of oxygen consumption by the mycelium of this strain. A much smaller respiration-suppressing effect was observed for mutant 1–6, for which the total inhibitory concentration of H_2O_2 was as high as 5% (Figure 7B). This finding shows that biotransformation of volatile terpenes can be run under unconventional oxygenation of the culture with hydrogen peroxide (H_2O_2) as a supplemental oxygen source [67,68].

Table 2. Comparison of resistance of the parental strain and mutant 1–6 of the fungus *C. pannorum* A-1 to hydrogen peroxide as measured by their H_2O_2 decomposition capacity.

Presence of Hydrogen Peroxide in the Medium After 1-h Incubation		
H_2O_2 Concentration % (*v/v*)	**Wild-Type Strain**	**Mutant 1–6**
Control	–	–
0.1	–	–
0.5	+	–
1.0	+	–
1.5	+	+
2.5	+	+
5.0	+	+

Control – mycelium incubated without an addition of H_2O_2; (–) – not detected; (+) – detected.

We showed that the mutants characterized by a higher GMA exhibited both a higher biotransformation rate and a higher resistance to H_2O_2-induced oxidative stress, compared to the wild strain; cell adaptation to monoterpene, β-pinene enhanced this effect. Sequential adaptation of *P. digitatum* NRRL 1202 to small doses of (R)-(+)-limonene led to a 12-fold increase in the efficiency of the biotransformation of this compound to α-terpineol in relation to an uninduced strain [69]. Adaptation of *Rhodococcus erythropolis* cells to increasing concentrations of carveol (substrate) and carvone (product) in n-dodecane prior to biotransformation resulted in an 8.3-fold increase in the carvone production rate [70]. Induction with styrene of *Pseudomonas* sp. VLB120DC, a strain capable of processing styrene into pure enantiomeric (S)-styrene oxide, allowed to obtain a specific oxygenase activity of 60–70 U/g cell dry weight (CDW), where the activity of uninduced cells remained at the level of 0.7 U/g CDW [42]. *Salmonella enterica* serovar Enteritidis adapted to ethanol developed cross-tolerance to malic acid, resulting in a doubling of the minimum bactericidal concentration and increased cell survival rates [71].

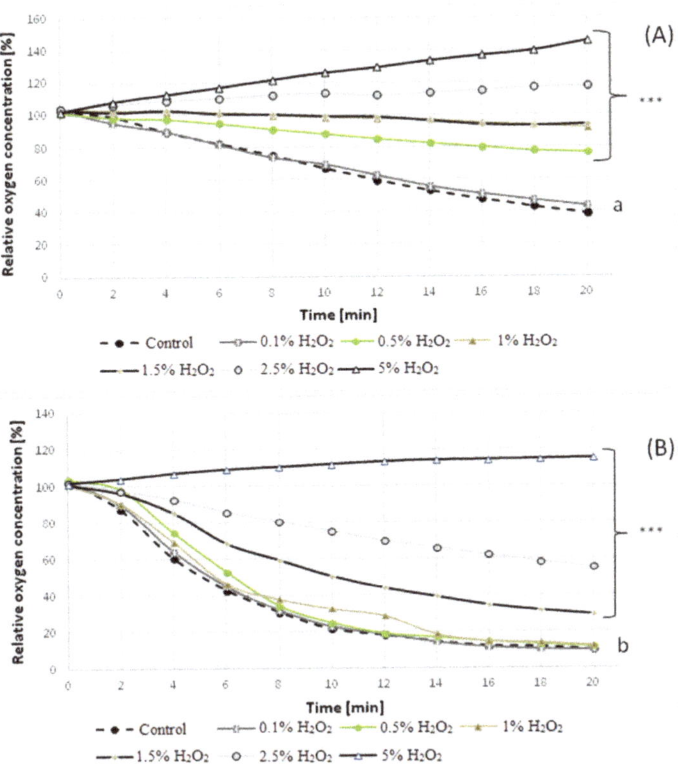

Figure 7. Oxygen consumption rate by the mycelium of the wild strain (**A**) and mutant 1–6 (**B**) of *C. pannorum* A-1 pre-incubated with different concentrations of H_2O_2. ***($p < 0.0001$) indicate statistically significant differences with respect to the control (mycelium incubated without the addition of H_2O_2). Lowercase letters (a, b) indicate that the differences between control wild strain and the 1–6 mutant are significant ($p \leq 0.0001$).

3. Materials and Methods

3.1. Chemicals

(1S)-(−)-β-Pinene (98%), N-methyl-N′-nitro-N-nitrosoguanidine (NTG) (97%), and *trans*-pinocarveol (≥96%) were purchased from Sigma-Aldrich, St Louis, MO, USA. (−)-Linalool, at a purity of >97%, was obtained from Fluka, Buchs, Switzerland, and was stored at 4 °C. Hydrogen peroxide (30%) and potassium iodide (>99%) were obtained from POCH, Gliwice, Poland.

3.2. Fungus and Media

The experiments were performed using the psychrotrophic fungus *C. pannorum* A-1 as the host strain. Fungal cells had been isolated from soil samples collected in the Arctic tundra (West

Spitsbergen) [72]. Prior to the experiments, the cells had been maintained on malt agar slants at 4 °C, and subcultured every month. The cultivation and bioconversion procedures were carried out using liquid basal medium (BM) composed of glucose 1%, malt extract 1%, peptone 0.5%, and yeast extract 0.5% (pH = 6.14). Adaptation experiments were carried out in adaptive medium M1 composed of agarized BM diluted 10-fold with distilled water with an addition of β-pinene at a concentration of 1% (*v/v*). β-pinene-resistant mutants were selected and isolated on the following solid media: 1. a gradient medium consisting of BM and water with 1% β-pinene (M2); 2. a minimal medium composed of water, mineral salts [K_2HPO_4 0.1%, $(NH_2)_2SO_4$ 0.5%, $MgSO_4$ 0.02%], and 1% β-pinene (M3); β-pinene was dispersed as an emulsion in 10-fold diluted BM using 0.01% Tween 80. Agarized BM was used to determine conidial survival.

3.3. Induction, Adaptation, and Selection of Mutants

Mutants were induced by treating a conidial suspension (2×10^6 spores mL^{-1}) with UV irradiation for 10 or 15 min (960 erg/mm^2) and NTG at a final concentration of 0.01% for 5 to 20 min (UV-NTG method) [73] with magnetic stirring (150 rpm) under room temperature according to the program given in Table S1. After mutagenization, the surviving spores were spread on dishes with agarized BM and incubated 4 days at 20 °C. The number of colonies formed was compared to control plates with non-treated spores. At this stage, the drop in viability was over 90%. For further experiments, random mutants which grew well in the presence of β-pinene were isolated. For this purpose, the colonies were passaged several times on solid BM with the addition of 1% β-pinene and then sown on M2 medium with a linear gradient of β-pinene concentrations according to the Gradient Plate technique [74].

To confirm the linear concentration of β-pinene in the agar plates, pieces of M2 medium were extracted and analyzed using GC. Agar discs (6 × 8 in diameter) were cut with a sterile cork borer from at least five regions. They were then extracted with EtOAc and subjected to pinene quantification.

For initial adaptation, a fragment of the mycelium of each mutant was transferred onto a Petri dish containing 10-fold diluted BM (M1) with a hollow in the center filled with β-pinene (50 μL). Colonies which grew the best were picked, and their general metabolic activity (GMA) was determined as the oxygen consumption rate in phosphate buffer according to the procedure described below. The most metabolically active mutants were transferred onto adaptive medium M1 containing β-pinene (adapted mutants) or onto wort medium M1 without the substrate (non-adapted mutants). The mutants were incubated for a month, after which another selection round was carried out based on GMA.

Mutants characterized by the highest GMA were transplanted onto slants with adaptive medium M1, stored at 4 °C, and subcultured every month.

3.4. General Metabolic Activity (GMA) Assay

The general metabolic activity of the mycelium was determined as the oxygen uptake rate using a calibrated VisiFerm DO Arc 225 electrode (Hamilton Robotics, Bonaduz, GR, Switzerland). Shaking flask cultures (150 rpm) of the mutants were grown in liquid BM at 20 °C. After two days of growth, 25 mL mycelial culture of *C. pannorum* A-1 containing a fixed amount of fungal mycelium (5 g of dry weight per L) was aseptically transferred into 100 mL beakers containing 50 mL of 0.05 M McIlvaine buffer (pH 6) with an addition of 0.5% glucose. Finally, the samples were aerated to a dissolved oxygen (DO) concentration of 100% air saturation and incubated with magnetic stirring (150 rpm) under a controlled temperature (20 °C). The electrode was calibrated before each experiment by measuring its signal in the air-saturated McIlvaine buffer prior to the addition of the mycelium. The values of the readings were expressed as percent of dissolved oxygen saturation and were then converted to oxygen uptake rate k [$\% \times s^{-1}$]. In all experiments, mycelium was sampled directly from the pellet culture. Care was taken to obtain samples representative of the different pellet sizes and mycelium weights. Three independent experiments were performed. The results are reported as means of all replicates.

3.5. Operation at Shake Flask Scale

To perform biotransformation experiments, 25 mL of sterile basal medium (in 100-mL Erlenmeyer flasks) was uniformly inoculated with 1.5 mL of *C. pannorum* spore suspension (at concentration of 2×10^6 mL^{-1}) and cultivated at 20 °C on a rotary shaker (150 rpm). The fungal spore suspension was prepared as an inoculum using 6-day old spores pre-grown at 20 °C on agar slants, which were then harvested, filtered through glass wool to remove hyphal fragments, and washed twice with sterile distilled water with 0.01% Tween 80, pH 5.0. Biotransformation was started by directly adding 1% *v/v* of β-pinene to the pre-grown 2 day-old mycelial culture using the procedure described in [43]. All β-pinene biotransformations were carried out in parallel with controls, in identical conditions, using heat-inactivated microorganisms which had been autoclaved for 15 min at 121 °C.

For the kinetic experiment, mycelia of the parental strain and mutant 1-6 previously cultured for 2 d using the procedure described above, were harvested washed twice with sterile 0.1 M McIlvaine buffer, pH 5.0. Next, equal amounts (by weight) of biomass were aseptically transferred to Erlenmeyer flasks containing 25 mL of sterile media. Biotransformation of β-pinene (1% (*v/v*)) was performed on BM and on phosphate buffer, pH 6, containing 1% glucose, using 1.5 and 2.35 g of dry mass per L, respectively. The entire contents of three flasks were sacrificed each time at regular intervals (0, 12, 24, 36, 48, 60, 72, and 96 h) to assay the products of biotransformation. Growth was followed by measuring the dry weight of mycelium. Mycelial biomass was determined gravimetrically, after washing with 0.1 M phosphate buffer, by drying to a constant weight in an oven at 105 °C for 3 h. It was expressed as g/L dry weight.

The biotransformation yield was defined as the quantity of the product per gram of dry weight of the mycelium per liter aqueous phase (mg/g d.w. × L).

3.6. Effect of Hydrogen Peroxide Concentration on the Metabolic Activity of the Parental and the Mutant Strain of C. pannorum A-1

Identical amounts (2.65 g of dry weight per L) of two-day-old mycelia of the parental strain and mutant 1-6 were washed twice with sterile 0.1 M phosphate buffer and transferred to 100 mL Erlenmeyer flasks containing 40 mL dissolved BM with an addition of hydrogen peroxide at concentrations of 0.1, 0.5, 1.0, 1.5, 2.5, and 5% (*v/v*). After 60-min incubation at 20 °C with magnetic stirring (150 rpm), the residual content of H_2O_2 was estimated in 1 mL aliquots of the medium using potassium iodide and starch solution as an indicator. The pre-incubated biomasses were filtrated, washed twice with sterile 0.9% NaCl solution, and then fixed amounts of the mycelia were suspended uniformly in 0.05 M McIlvaine buffer (pH 6) with an addition of 1% glucose in order to measure the GMA of living mycelia, according to procedure described in the 'General metabolic activity assay'.

3.7. Biotransformation Analysis

After the specified biotransformation time, the biomass was harvested by filtration, and the liquid for product recovery was extracted twice with an equal volume of diethyl ether in a separation funnel for 10 min. Before extraction, 250 µL of a 0.1% internal standards (IS) solution in hexane was added to the filtrate. The ether fraction was collected and concentrated on rotary vacuum evaporators (BUCHI, Rotavapor R-200/205, Flawil, Schwitzerland) at a water bath temperature of 40 °C under mild pressure of 800 mbar. The residues obtained were dissolved in 6 mL of hexane. Gas chromatography with flame-ionization detector (GC-FID) (VARIAN, Palo Alto, California, USA) and mass specta coupled to gas chromatography (GC-MS) (Thermo Finnigan, Trace DSQ GC/MS Ultra Chromatograph, Austin, TX, USA) analyses of terpenes were conducted using the method reported previously [43]. Peaks were identified by fitting the mass spectra to those from standard library database systems (HP Wiley, NIST) and the MassFinder library and by comparing the GC retention indices to those of standard compounds. Biotransformations were performed in three replicate samples, and the analyses were carried out in duplicate. The error associated with the GC quantification of the samples was ±6% and was quoted for a confidence interval of 94%. The data are reported as mean values.

3.8. Statistical Analysis

The experiments were carried out in triplicate. The results shown in Figures 1 and 4–6 were analyzed by a one-way analysis of variance (one-way ANOVA) followed by Tukey's post-hoc test for multiple comparisons. Data from the comparison of the control strain and the 1–6 mutant were analyzed statistically using the unpaired Student's t-test (Figure 7). The statistical analysis was carried out using GraphPad Prism for Windows version 5.03 (GraphPad Software, San Diego, CA, USA). All results are presented as mean +/- SEM. $p < 0.05$ was considered statistically significant with a confidence level of 95%.

4. Conclusions

The present results demonstrate that the adapted mutants of *C. pannorum* A-1 are characterized by an enhanced resistance to stressful conditions. So far, however, they have shown moderate biocatalytic activities, giving relatively high *trans*-pinocarveol yields, which means they are still subject to optimization. The maximum *trans*-pinocarveol concentrations obtained in the experiment reported in the present paper were in the range of 147.2–314.7 mg/g d.w. × L. In our non-optimized flask system, we were able to biotransform of β-pinene worth US$ 0.36 to *trans*-pinocarveol with a value of US$ 67 (per 1 L of batch culture), using the most active mutant. This result indicates that *C. pannorum* A-1 and its mutants are efficient biocatalysts for the conversion of β-pinene to this valuable product. Further improvement in β-pinene biotransformation may be achieved by selecting blocked mutants of the isolated fungal strains, which would prevent further metabolization of the main product (*trans*-pinocarveol). Filtration-enrichment methods for selecting auxotrophs may be used for this purpose [75]. Moreover, optimization of the culture conditions in bioreactors is necessary for the strain to be utilized in larger scale bioconversion processes, as some parameters, such as oxygen supply or pH, cannot be controlled in a flask culture.

The results show that it is possible to increase the rate of positive mutations by using media with a gradient of the toxic substrate and performing rapid initial evaluation of the GMA of the examined mutants of *C. pannorum* A-1. The use of the gradient plate method and subsequent determination of the GMA of the isolated mutants greatly simplifies the selection of substrate-resistant strains. Direct screening has the advantage of significantly reducing the number of cultures isolated from plates, which would normally require productivity to be tested in shake flask cultures. This method may considerably improve the selection of β-pinene biotransformation-active microbial strains and also other cultures showing biotransformation activity toward toxic substrates.

Supplementary Materials: The following are available online, Figure S1: Agar plate with linear gradient of β-pinene concentration, Figure S2: Flowchart of the procedure for improving *C. pannorum* A-1 as a biocatalyst for β-pinene biotransformation, Figure S3: Mass spectra of unknown compounds; RI = 1275, RT = 20.1 min, Table S1: Variants of mutagenesis of the psychrotrophic fungus *C. pannorum* A-1 and their impact on survivability of conidia. Survivability was expressed as the number of colonies formed compared to control plates with non-treated spores after 2 days of incubation at 20 °C on agarized BM, Table S2: Oxygen uptake rate k [$\% \times s^{-1}$] for the 12 most active GMA mutants and parental strain. The standard deviation was approximately 3%

Author Contributions: Conceptualization, J.F. and M.T.; methodology, M.K., J.F., A.G. and M.T.; software, M.K., J.F., K.J. and M.T.; validation, M.K., A.G. and M.T.; formal analysis, M.K., K.J. and M.T.; investigation, M.K., K.J. and A.G.; resources, M.K., J.F., A.G., and M.T.; data curation, M.K. and M.T.; writing—original draft preparation, J.F. and M.T.; writing—review and editing, M.K., J.F. and M.T.; visualization, M.K. and K.J.; supervision, J.F. and M.T.; project administration, J.F.; funding acquisition, M.K, J.F., A.G., K.J. and M.T. All authors have read and agreed to the published version of the manuscript.

Funding: This research received no external funding.

Acknowledgments: The authors would like to thank Maria Curie-Skłodowska University in Lublin, Poland, for providing institutional funds to support this work.

Conflicts of Interest: The authors declare no conflict of interest.

References

1. Vespermann, K.A.C.; Paulino, B.N.; Barcelos, M.C.S.; Pessôa, M.G.; Pastore, G.M.; Molina, G. Biotransformation of α- and β-pinene into flavor compounds. *Appl. Microbiol. Biotechnol.* **2017**, *101*, 1805–1817. [CrossRef] [PubMed]
2. Surburg, H.; Panten, J. *Common Fragrance and Flavor Materials: Preparation, Properties and Uses, Completely Revised and Enlarged Edition*, 5th ed.; Wiley-VCH Verlag GmbH: Weinheim, Germany, 2006.
3. Paduch, R.; Kandefer-Szerszeń, M.; Trytek, M.; Fiedurek, J. Terpenes: Substances useful in human healthcare. *Arch. Immunol. Ther. Exp.* **2007**, *55*, 315–327. [CrossRef] [PubMed]
4. Trytek, M.; Paduch, R.; Pięt, M.; Kozieł, A.; Kandefer-Szerszeń, M.; Szajnecki, Ł.; Gromada, A. Biological activity of oxygenated pinene derivatives on human colon normal and carcinoma cells. *Flavour Fragr. J.* **2018**, *33*, 428–437. [CrossRef]
5. Viana, A.F.S.C.; Lopes, M.T.P.; Oliveira, F.T.B.; Nunes, P.I.G.; Santos, V.G.; Braga, A.D.; Silva, A.C.A.; Sousa, D.P.; Viana, D.A.; Rao, V.S.; et al. (-)-Myrtenol accelerates healing of acetic-induced gastric ulcers in rats and human gastric adenocarcinoma cells. *Eur. J. Pharmacol.* **2019**, *854*, 139–148. [CrossRef] [PubMed]
6. De Cássia da Silveira e Sá, R.; Andrade, L.N.; De Sousa, D.P. A Review on anti-inflammatory activity of monoterpenes. *Molecules* **2013**, *18*, 1227–1254. [CrossRef]
7. Hollmann, F.; Arends, I.W.; Buehler, K.; Schallmey, A.; Bühler, B. Enzyme-mediated oxidations for the chemist. *Green Chem.* **2011**, *13*, 226–265. [CrossRef]
8. Sales, A.; Paulino, B.N.; Pastore, G.M.; Bicas, J.L. Biogeneration of aroma compounds. *Curr. Opin. Food Sci.* **2018**, *19*, 77–84. [CrossRef]
9. Molina, G.; Pessôa, M.G.; Bicas, J.L.; Fontanille, P.; Larroche, C.; Pastore, G.M. Optimization of limonene biotransformation for the production of bulk amounts of α-terpineol. *Bioresour. Technol.* **2019**, *294*, 122180. [CrossRef]
10. Molbase: Chemical B2B E-Commerce Platform. 2018. Available online: https://www.molbase.com (accessed on 3 October 2018).
11. De Oliveira Felipe, L.; De Oliveira, A.M.; Bicas, J.L. Bioaromas—Perspectives for sustainable development. *Trends Food Sci. Technol.* **2017**, *62*, 141–153. [CrossRef]
12. Gavrilov, V.V.; Startseva, V.A.; Nikitina, L.E.; Lodochnikova, O.A.; Gnezdilov, O.I.; Lisovskaya, S.A.; Clushko, N.I.; Klimovitskii, E.N. Synthesis and antifungal activity of sulfides, sulfoxides, and sulfones based on (1S)-(-)-β-pinene. *Pharm. Chem. J.* **2010**, *44*, 126–129. [CrossRef]
13. Silva, A.C.R.; Lopes, P.M.; Azevedo, M.M.B.; Costa, D.C.M.; Alviano, C.S.; Alviano, D.S. Biological activities of α-pinene and β-pinene enantiomers. *Molecules* **2012**, *17*, 6305–6316. [CrossRef] [PubMed]
14. Guzman-Gutierrez, S.L.; Gomez-Cansino, R.; Garcia-Zebadua, J.C.; Jimenez-Perez, N.C.; Reyes-Chilpa, R. Antidepressant activity of *Litsea glaucescens* essential oil: Identification of β-pinene and linalool as active principles. *J. Ethnopharmacol.* **2012**, *143*, 673–679. [CrossRef] [PubMed]
15. Guzman-Gutierrez, S.L.; Bollina-Jaime, H.; Gomez-Cansino, R.; Reyes-Chilpa, R. Linalool and β-pinene exert their antidepressant-like activity through the monoaminergic pathway. *Life Sci.* **2015**, *128*, 24–29. [CrossRef] [PubMed]
16. Bajpai, V.K.; Rahman, A.; Choi, U.K.; Youn, S.J.; Kang, S.C. Inhibitory parameters of the essential oil and various extracts of *Metasequoia glyptostroboides Miki ex Hu* to reduce food spoilage and food-borne pathogens. *Food Chem.* **2007**, *105*, 1061–1066. [CrossRef]
17. Tavares, C.S.; Martins, A.; Faleiro, M.L.; Miguel, M.M.; Duarte, L.C.; Gameiro, J.A.; Roseiro, L.B.; Figueiredo, A.C. Bioproducts from forest biomass: Essential oils and hydrolates from wastes of *Cupressus lusitanica* Mill. and *Cistus ladanifer* L. *Ind. Crop. Prod.* **2020**, *144*, 112034. [CrossRef]
18. Drosopoulou, E.; Vlastos, D.; Efthimiou, I.; Kyrizaki, P.; Tsamadou, S.; Anagnostopoulou, M.; Kofidou, D.; Gavriilidis, M.; Mademtzoglou, D.; Mavragani-Tsipidou, P. In vitro and in vivo evaluation of the genotoxic and antigenetoxic potential of the major Chios mastic water constituents. *Sci. Rep.* **2018**, *8*, 12200. [CrossRef]
19. Rozenbaum, H.F.; Patitucci, M.L.; Antunes, O.A.C.; Pereira, N. Production of aromas and fragrances through microbial oxidation of monoterpenes. *Braz. J. Chem. Eng.* **2006**, *23*, 273–279. [CrossRef]
20. Lee, S.Y.; Kim, S.H.; Hong, C.Y.; Kim, H.Y.; Ryu, S.H.; Choi, I.G. Biotransformation of (-)-α-pinene by whole cells of white rot fungi, *Ceriporia* sp. ZLY-2010 and *Stereum hirsutum*. *Mycobiology* **2015**, *43*, 297–302. [CrossRef]

21. Wang, J.; Li, R.; Lu, D.; Ma, S.; Yan, Y.; Li, W. A quick isolation method for mutants with high lipid yield in oleaginous yeast. *World J. Microbiol. Biotechnol.* **2009**, *25*, 921–925. [CrossRef]
22. Harper, M.; Lee, C.J. Genome-wide analysis of mutagenesis bias and context sensitivity of N-methyl-N-nitro-N-nitrosoguanidine (NTG). *Mutat. Res.* **2012**, *731*, 64–67. [CrossRef]
23. Wakai, S.; Arazoe, T.; Ogino, C.; Kondo, A. Future insights in fungal metabolic engineering. *Bioresour. Technol.* **2017**, *245*, 1314–1326. [CrossRef] [PubMed]
24. Meyer, V. Genetic engineering of filamentous fungi—Progress, obstacles and future trends. *Biotechnol. Adv.* **2008**, *26*, 177–185. [CrossRef]
25. Nakane, A.; Takamatsu, H.; Oguro, A.; Sodaie, Y.; Nakamura, K.; Yamane, K. Acquisition of azide-resistance by elevated SecA ATPase activity confers azide-resistance upon cell growth and protein translocation in *Bacillus subtilis*. *Microbiology* **1995**, *141*, 113–121. [CrossRef] [PubMed]
26. Skowronek, M.; Fiedurek, J. Selection of biochemical mutants of *Aspergillus niger* resistant to some abiotic stresses with increased inulinase production. *J. Appl. Microbiol.* **2003**, *95*, 686–692. [CrossRef]
27. Fiedurek, J.; Trytek, M.; Szczodrak, J. Strain improvement of industrially important microorganisms based on resistance to toxic metabolites and abiotic stresses. *J. Basic Microbiol.* **2017**, *57*, 445–459. [CrossRef] [PubMed]
28. Fiedurek, J.; Trytek, M.; Skowronek, M. Strategies for improving the efficiency of bioprocesses involving toxic metabolites. *Curr. Org. Chem.* **2012**, *16*, 2946–2960. [CrossRef]
29. Kirimura, K.; Sarangbin, S.; Rugsaseel, S.; Usami, S. Citric acid production by 2-deoxyglucose-resistant mutant strains of *Aspergillus niger*. *Appl. Microbiol. Biotechnol.* **1992**, *36*, 573–577. [CrossRef]
30. Aono, R.; Inoue, A. Organic solvent tolerance in microorganisms. In *Microbial Life in Extreme Environments;* Horikoshi, K., Grant, W.D., Eds.; Wiley-Liss Inc.: New York, NY, USA, 1998; pp. 287–310.
31. Meyer, D.; Buehler, B.; Schmid, A. Process and catalyst design objectives for specific redox biocatalysis. *Adv. Appl. Microbiol.* **2006**, *59*, 53–91.
32. Doukyu, N.; Ishikawa, K.; Watanabe, R.; Ogino, H. Improvement in organic solvent tolerance by double disruptions of proV and marR genes in *Escherichia coli*. *J. Appl. Microbiol.* **2012**, *112*, 464–474. [CrossRef]
33. Asako, H.; Nakajima, H.; Kobayashi, K.; Kobayashi, M. Organic solvent tolerance and antibiotic resistance increased by overexpression of marA in *Escherichia coli*. *Appl. Environ. Microbiol.* **1997**, *63*, 1428–1433. [CrossRef]
34. Volmer, J.; Neumann, C.; Bühler, B.; Schmid, A. Engineering of *Pseudomonas taiwanensis* VLB120 for constitutive solvent tolerance and increased specific styrene epoxidation activity. *Appl. Environ. Microbiol.* **2014**, *80*, 6539–6548. [CrossRef] [PubMed]
35. Nishida, N.; Ozato, N.; Matsui, K.; Kuroda, K. ABC transporters and cell wall proteins involved in organic solvent tolerance in *Saccharomyces cerevisiae*. *J. Biotechnol.* **2013**, *165*, 145–152. [CrossRef] [PubMed]
36. Shah, M.K.; Bergholz, T.M. Variation in growth and evaluation of cross-protection in *Listeria monocytogenes* under salt and bile stress. *J. Appl. Microbiol.* **2020**. [CrossRef] [PubMed]
37. Tai, Y.N.; Xu, M.; Ren, J.N.; Dong, M.; Yang, Z.Y.; Pan, S.Y.; Fan, G. Optimisation of α-terpineol production by limonene biotransformation using *Penicillium digitatum* DSM 62840. *J. Sci. Food Agric.* **2016**, *96*, 954–961. [CrossRef]
38. Rottava, I.; Cortina, P.F.; Grando, C.E.; Colla, A.R.S.; Martello, E.; Cansian, R.L.; Toniazzo, G.; Treichel, H.; Antunes, O.A.C.; Oestreicher, E.G.; et al. Isolation and screening of microorganisms for R-(+)-limonene and (−)-β-pinene biotransformation. *Appl. Biochem. Biotechnol.* **2010**, *162*, 719–732. [CrossRef]
39. Santos, P.M.; Sá-Correia, I. Adaptation to β-myrcene catabolism in *Pseudomonas* sp. M1: An expression proteomics analysis. *Proteomics* **2009**, *9*, 5101–5111. [CrossRef]
40. Kakuk, B.; Kovács, K.L.; Szuhaj, M.; Rákhely, G.; Bagi, Z. Adaptation of continuous biogas reactors operating under wet fermentation conditions to dry conditions with corn stover as substrate. *Anaerobe* **2017**, *46*, 78–85. [CrossRef]
41. Van Hamme, J.D.; Singh, A.; Ward, O.P. Recent advances in petroleum microbiology. *Microbiol. Mol. Biol. Rev.* **2003**, *67*, 503–549. [CrossRef]
42. Park, J.B.; Bühler, B.; Panke, S.; Witholt, B.; Schmid, A. Carbon metabolism and product inhibition determine the epoxidation efficiency of solvent-tolerant *Pseudomonas* sp. strain VLB120DC. *Biotechnol. Bioeng.* **2007**, *98*, 1219–1229. [CrossRef]
43. Trytek, M.; Jędrzejewski, K.; Fiedurek, J. Bioconversion of α-pinene by a novel cold-adapted fungus *Chrysosporium pannorum*. *J. Ind. Microbiol. Biotechnol.* **2015**, *42*, 181–188. [CrossRef]

44. Trytek, M.; Fiedurek, J.; Gromada, A. Effect of some abiotic stresses on the biotransformation of α-pinene by a psychrotrophic *Chrysosporium pannorum*. *Biochem. Eng. J.* **2016**, *112*, 86–93. [CrossRef]
45. Agrawal, R.; Deepika, N.U.A.; Joseph, R. Strain improvement of *Aspergillus* sp. and *Penicillium* sp. by induced mutation for biotransformation of α-pinene to verbenol. *Biotechnol. Bioeng.* **1999**, *63*, 249–252. [CrossRef]
46. Lloyd, D.; Morrell, S.; Carlsen, H.N.; Degn, H.; James, P.E.; Rowlands, C.C. Effects of growth with ethanol on fermentation and membrane fluidity of *Saccharomyces cerevisiae*. *Yeast* **1993**, *9*, 825–833. [CrossRef] [PubMed]
47. Kabelitz, N.; Santos, P.M.; Heipieper, H.J. Effect of aliphatic alcohols on growth and degree of saturation of membrane lipids in *Acinetobacter calcoaceticus*. *FEMS Microbiol. Lett.* **2003**, *220*, 223–227. [CrossRef]
48. Neumann, G.; Kabelitz, N.; Zehnsdorf, A.; Miltner, A.; Lippold, H.; Meyer, D.; Schmid, A.; Heipieper, H.J. Prediction of the adaptability of *Pseudomonas putida* DOT-T1E to a second phase of a solvent for economically sound two-phase biotransformations. *Appl. Environ. Microbiol.* **2005**, *71*, 6606–6612. [CrossRef] [PubMed]
49. Ortiz-Zamora, O.; Cortes-García, R.; Ramírez-Lepe, M.; Gómez-Rodríguez, J.; Aguilar-Uscanga, M.G. Isolation and selection of ethanol-resistant and osmotolerant yeasts from regional agricultural sources in Mexico. *J. Food Process. Eng.* **2008**, *32*, 775–786. [CrossRef]
50. Loffler, C.; Eberlein, C.; Mausezahl, I.; Kappelmeyer, U.; Heipieper, H.J. Physiological evidence for the presence of a cis-trans isomerase of unsaturated fatty acids in *Methylococcus capsulatus* Bath to adapt to the presence of toxic organic compounds. *FEMS Microbiol. Lett.* **2010**, *308*, 68–75. [CrossRef]
51. Naether, D.J.; Slawtschew, S.; Stasik, S.; Engel, M.; Olzog, M.; Wick, L.Y.; Timmis, K.N.; Heipieper, H.J. Adaptation of the hydrocarbonoclastic bacterium *Alcanivorax borkumensis* SK2 to alkanes and toxic organic compounds: A physiological and transcriptomic approach. *Appl. Environ. Microbiol.* **2013**, *79*, 4282–4293. [CrossRef]
52. Stanley, D.; Fraser, S.; Chambers, P.J.; Rogers, P.; Stanley, G.A. Generation and characterisation of stable ethanol-tolerant mutants of *Saccharomyces cerevisiae*. *J. Ind. Microbiol. Biotechnol.* **2010**, *37*, 139–149. [CrossRef]
53. Weber, F.J.; Hage, K.C.; De Bont, J.A.M. Growth of the fungus *Cladosporium sphaerospermum* with toluene as the sole carbon and energy source. *Appl. Environ. Microbiol.* **1995**, *61*, 3562–3566. [CrossRef]
54. Weber, F.J.; De Bont, J.A.M. Adaptation mechanisms of microorganisms to the toxic effects of organic solvents on membranes. *Biochim. Biophys. Acta* **1996**, *1286*, 225–245. [CrossRef]
55. Torres, S.; Pandey, A.; Castro, G.R. Organic solvent adaptation of Gram positive bacteria: Applications and biotechnological potentials. *Biotechnol. Adv.* **2011**, *29*, 442–452. [CrossRef] [PubMed]
56. Brandt, B.W.; Kelpin, F.D.L.; Van Leeuwen, I.M.M.; Kooijman, S.A.L.M. Modelling microbial adaptation to changing availability of substrates. *Water Res.* **2004**, *38*, 1003–1013. [CrossRef] [PubMed]
57. Bajaj, S.; Singh, D.K. Biodegradation of persistent organic pollutants in soil, water and pristine sites by cold-adapted microorganisms: Mini review. *Int. Biodeterior. Biodegrad.* **2015**, *100*, 98–105. [CrossRef]
58. Králová, S. Role of fatty acids in cold adaptation of Antarctic psychrophilic *Flavobacterium* spp. *Syst. Appl. Microbiol.* **2017**, *40*, 329–333. [CrossRef]
59. Brininger, C.; Spradlin, S.; Cobani, L.; Evilia, C. The more adaptive to change, the more likely you are to survive: Protein adaptation in extremophiles. *Sem. Cell. Dev. Biol.* **2018**, *84*, 158–169. [CrossRef]
60. Dungait, J.A.J.; Kemmitt, S.J.; Michallon, L.; Guo, S.; Wen, Q.; Brookes, P.C.; Evershed, R.P. The variable response of soil microorganisms to trace concentrations of low molecular weight organic substrates of increasing complexity. *Soil Biol. Biochem.* **2013**, *64*, 57–64. [CrossRef]
61. Wery, J.; De Bont, J.A.M. Solvent-tolerance of pseudomonads: A new degree of freedom in biocatalysis. In *Pseudomonas, Volume 3: Biosynthesis of Macromolecules and Molecular Metabolism*; Ramos, J.L., Ed.; Kluwer Academic; Plenum Publishers: New York, NY, USA, 2004; pp. 609–634.
62. Ramos, J.L.; Duque, E.; Gallegos, M.T.; Godoy, P.; Ramos-Gonzalez, M.I.; Rojas, A.; Teran, W.; Segura, A. Mechanisms of solvent tolerance in gram-negative bacteria. *Annu. Rev. Microbiol.* **2002**, *56*, 743–768. [CrossRef]
63. Heipieper, H.J.; Neumann, G.; Cornelissen, S.; Meinhardt, F. Solvent-tolerant bacteria for biotransformations in two-phase fermentation systems. *Appl. Microbiol. Biotechnol.* **2007**, *74*, 961–973. [CrossRef]
64. Fontanille, P.; Larroche, C. Optimization of isonovalal production from α-pinene oxide using permeabilized cells of *Pseudomonas rhodesiae* CIP 107491. *Appl. Microbiol. Biotechnol.* **2003**, *60*, 534–540. [CrossRef]
65. Alonso-Calleja, C.; Guerrero-Ramos, E.; Alonso-Hernando, A.; Capita, R. Adaptation and cross-adaptation of *Escherichia coli* ATCC 12806 to several food-grade biocides. *Food Control* **2015**, *56*, 86–94. [CrossRef]

66. Cotter, P.D.; Hill, C. Surviving the acid test: Responses of gram-positive bacteria to low pH. *Microbiol. Mol. Biol. Rev.* **2003**, *67*, 429–453. [CrossRef] [PubMed]
67. Fiorenza, S.; Ward, C.H. Microbial adaptation to hydrogen peroxide and biodegradation of aromatic hydrocarbons. *J. Ind. Microbiol. Biotechnol.* **1997**, *18*, 140–151. [CrossRef] [PubMed]
68. Trytek, M.; Fiedurek, J.; Skowronek, M. Biotransformation of (*R*)-(+)-limonene by the psychrotrophic fungus *Mortierella minutissima* in H_2O_2-oxygenated culture. *Food Technol. Biotechnol.* **2009**, *47*, 131–136.
69. Tan, Q.; Day, D.F.; Cadwallader, K.R. Bioconversion of (R)-(+)-limonene by *P. digitatum* (NRRL 1202). *Process Biochem.* **1998**, *33*, 29–31. [CrossRef]
70. De Carvalho, C.C.C.R.; Poretti, A.; Da Fonseca, M.M.R. Cell adaptation to solvent, substrate and product: A successful strategy to overcome product inhibition in a bioconversion system. *Appl. Microbiol. Biotechnol.* **2005**, *69*, 268–275. [CrossRef]
71. He, S.; Cui, Y.; Qin, X.; Zhang, F.; Shi, C.; Paoli, G.C.; Shi, X. Influence of ethanol adaptation on *Salmonella enteric* serovar *Enteritidis* survival in acidic environments and expression of acid tolerance-related genes. *Food Microbiol.* **2018**, *72*, 193–198. [CrossRef]
72. Fiedurek, J.; Gromada, A.; Słomka, A.; Korniłowicz-Kowalska, T.; Kurek, E.; Melke, J. Catalase activity in arctic microfungi grown at different temperatures. *Acta Biol. Hung.* **2003**, *54*, 105–112. [CrossRef]
73. Fiedurek, J.; Rogalski, J.; Ilczuk, Z.; Leonowicz, A. Screening and mutagenization of molds for the improvement of glucose oxidase production. *Enzym. Microb. Technol.* **1986**, *8*, 734–736. [CrossRef]
74. Bryson, V.; Szybalski, W. Microbial selection. *Science* **1952**, *116*, 45–51. [CrossRef]
75. Bos, C.J.; Debets, A.J.M.; Nachtegaal, H.; Slakhorst, S.M.; Swart, K. Isolation of auxotrophic mutants of *Aspergillus niger* by filtration enrichment and lytic enzymes. *Curr. Genet.* **1992**, *21*, 117–120. [CrossRef]

Sample Availability: Samples of the mutant strains are available from the authors.

MDPI
St. Alban-Anlage 66
4052 Basel
Switzerland
Tel. +41 61 683 77 34
Fax +41 61 302 89 18
www.mdpi.com

Molecules Editorial Office
E-mail: molecules@mdpi.com
www.mdpi.com/journal/molecules

www.ingramcontent.com/pod-product-compliance
Lightning Source LLC
LaVergne TN
LVHW070648170726
843515LV00004B/355
* 9 7 8 3 0 3 9 4 3 5 7 1 5 *